Power Systems

Electrical power has been the technological foundation of industrial societies for many years. Although the systems designed to provide and apply electrical energy have reached a high degree of maturity, unforeseen problems are constantly encountered, necessitating the design of more efficient and reliable systems based on novel technologies. The book series Power Systems is aimed at providing detailed, accurate and sound technical information about these new developments in electrical power engineering. It includes topics on power generation, storage and transmission as well as electrical machines. The monographs and advanced textbooks in this series address researchers, lecturers, industrial engineers and senior students in electrical engineering.

** Power Systems is indexed in Scopus**

More information about this series at http://www.springer.com/series/4622

Dharmesh Patel · Nilesh Chothani

Digital Protective Schemes for Power Transformer

Springer

Dharmesh Patel
Government Engineering College, Bharuch
Bharuch, Gujarat, India

Nilesh Chothani
Adani Institute of Infrastructure Engineering
Ahmedabad, Gujarat, India

ISSN 1612-1287 ISSN 1860-4676 (electronic)
Power Systems
ISBN 978-981-15-6765-0 ISBN 978-981-15-6763-6 (eBook)
https://doi.org/10.1007/978-981-15-6763-6

This Springer imprint is published by the registered company Springer Nature Singapore Pte Ltd.
The registered company address is: 152 Beach Road, #21-01/04 Gateway East, Singapore 189721, Singapore

Preface

The authors are happy to present this book to the readers of various levels. This book contains various protective techniques that have been proposed by the authors for transformer protection. This book helps researchers to understand various machine learning and digital techniques that have been utilized here for transformer protection.

The whole book manuscript has been organized into seven chapters as follows.

Chapter 1 outlines the motivation, problem statements and objectives with an introduction of traditional protection philosophy adapted to shelter the power system under consideration along with the state-of-the-art reviews on the existing methods. The literature review starts with the technological developments in the field of phasor estimation of an analog input signal applied to numerical relays to initiate the relaying actions. It also covers the reviews on numerical differential protection schemes along with a deep review on widely used methods based on adaptive digital differential protection, DFT/FFT and other filtration-based analysis, artificial intelligence-based, wavelet transform technique and SVM-based techniques. This chapter also covers an exhaustive literature survey on transformer protections against abnormal conditions.

Chapter 2 reveals the current transformer (CT) saturation detection and compensation algorithm in a power system with considering various effects. MDFT-based compensating algorithm has also been proposed to reconstruct the saturated samples. The proposed algorithm depends on a saturation detection index which is derived using derivatives of current signals and Newton's backward difference formulas. Validation of the proposed scheme is also carried out on a developed laboratory prototype. A comparative evaluation of the proposed algorithm is also carried out with existing schemes. Series of test results from simulation software and laboratory prototype show the effectiveness of the proposed CT saturation detection scheme.

Chapter 3 presents critical issues that influence the performance of the numerical percentage bias differential relays along with appropriate mathematical fundamentals. This chapter includes inrush detection with second-order derivative of differential current. It also comprises phasor angle comparison-based internal/external fault discriminations along with percentage biased differential protective scheme. FCDFT algorithm is implemented to validate the differential protective scheme on both PSCAD™ simulation and laboratory prototype.

Chapter 4 demonstrates an adaptive concept of the differential characteristic employed in the algorithm to maintain the stability of relay during external fault with CT saturation. It proposes an innovative solution over conventionally used relaying schemes. The prototype result on 2 kVA, 230/110V, single-phase transformer shows that the proposed scheme is capable to discriminate inrush, internal and external fault also with CT saturation conditions.

Chapter 5 outlines a novel scheme, based on relevance vector machine (RVM) as a fault classifier. RVM-based classifier discriminates various internal faults and abnormal conditions within a short time and having high accuracy up to 99% compared to SVM- and PNN-based classifier techniques. Power system is simulated in PSCAD™ software, and algorithm is validated through MATLAB software. The result in terms of fault classification accuracy and time shows the effectiveness of the presented protection scheme.

Chapter 6 discloses a new hierarchical ensemble extreme learning machine (HE-ELM)-based classifier technique to identify faults in and out of the transformer. The component ELM is structured hierarchically to improve its fault data classification accuracy. The developed algorithm is evaluated by PSCAD software and also successfully tested on hardware prototype in a laboratory environment. Results demonstrate that HE-ELM outperforms than existing schemes in the cross-domain recognition task.

Chapter 7 exhibits electrical and non-electrical parameter-based power transformer monitoring and protection. Various data such as core flux, age of the asset, heat generation, current harmonics and temperature are monitored in real time and processed it accordingly to enhance the working capability of the transformer. The proposed scheme is successfully tested on laboratory, and a fitness function is estimated from the collected data to examine the working condition of the transformer. Moreover, voltage, current and power-based inrush detection, as well as adaptive power differential protection (APDP), are applied to protect the transformer against fault. The hardware implementation and result validation prove the effectiveness of the proposed scheme to enhance the reliability of the grid which contains distribution transformer.

At the end, the conclusion and future scope are elaborated in detail. Details of simulation and hardware parameters are given in an appendix. Literatures used during the preparation of book are outlined in reference section.

Bharuch, Gujarat Dharmesh Patel
Ahmedabad, Gujarat Nilesh Chothani

Acknowledgements

This book is based on the research work carried out towards the digital revolution in transformer protection. We are grateful to the Government of India for allotted funds towards the research. The financial support is provided by the Science and Engineering Research Board (SERB) under the Department of Science and Technology (DST), India, with project ref. no. EMR/2016/006041.

We are grateful to the following journals for permission to reprint essays: Chap. 2 was published as "New Algorithm for Current Transformer Saturation Detection and Compensation Based on Derivatives of Secondary Currents and Newton's Backward Difference Formulae", IET Generation Transmission and Distribution, 8 (2014): 841–850; Chap. 3 was published as "Discrimination of Inrush, Internal, and External Fault in Power Transformer Using Phasor Angle Comparison and Biased Differential Principle", Electrical Power Components and Systems, 46 (2018): 788–801; Chap. 4 was published as "Adaptive Algorithm for Distribution Transformer Protection to Improve Smart Grid Stability", International Journal of Emerging Electric Power Systems, 19 (2018): 1–14; Chap. 5 was published as "Design and Development of Fault Classification Algorithm Based on Relevance Vector Machine for Power Transformer", IET Electrical Power Applications, 12 (2018): 557–565; Chap. 6 was published as "Identification of Internal Fault against External Abnormalities in Power Transformer Using Hierarchical Ensemble Extreme Learning Machine (HE-ELM) Technique", IET Science, Measurement and Technology, 14 (2020): 111–121; Chap. 7 was published as "Real-Time Monitoring and Adaptive Protection of Power Transformer to Enhance Smart Grid Reliability", Journal of Electrical Control and Communication Engineering, 15 (2019): 104–112.

We are expressing our sincere thanks to Sardar Vallabhbhai National Institute of Technology (SVNIT), Surat, Gujarat, India, and A. D. Patel Institute of Technology (ADIT), V. V. Nagar, Anand, Gujarat, India, for providing constant support in the execution of the work presented in this book. Moreover, we are also grateful to the staff members of these institutes for their continuous support.

We extend our special thanks to Dr. Bhavesh Bhalja, Associate Professor, IIT Roorkee, for his continuous guidance and encouragement. We are also deeply thankful to Dr. Khyati Mistry, Associate Professor, SVNIT, Surat, and Mr. Maulik Raichura, Research Scholar, Gujarat Technological University, for their interactions on the application and implementation of the suggested digital protection technique in laboratory.

Nobody has been more important to us in the pursuit of this book project than the members of our family. We would like to thank our family members for moral support, motivation and guidance to complete this monograph. We would like to thank all of them who have supported directly or indirectly from all the aspects towards the completion of this book project. Further, we are expressing deepest gratitude to the supreme power for helping us during every moment to complete this book.

Special thanks to the Springer Nature publication and associated press for the care they have given during the preparation and production of this book.

Contents

1 Introduction to Power Transformer Protection 1
1.1 Introduction 1
1.2 Types of Faults and Abnormalities 3
1.2.1 Internal Fault 3
1.3 External Fault or Abnormalities 4
1.4 Various Protective Schemes Used in Power Transformers 5
1.4.1 Over Current Protection 5
1.4.2 Overcurrent Protection with Harmonic Restraint Unit (HRU) 6
1.4.3 Restricted Earth Fault (REF) 6
1.4.4 Differential Protection 7
1.5 Burning Issues for Transformer Protection 9
1.5.1 Magnetizing Inrush Phenomenon 9
1.5.2 Current Transformer Saturation Conditions 11
1.5.3 Over Fluxing Condition 11
1.5.4 Inter-turn Fault Protection 12
1.6 Non-electrical Protection 12
1.6.1 Thermal Relays 12
1.6.2 Buchholz Relay 13
1.6.3 Sudden Pressure Relay 14
1.7 Overall Arrangements of Transformer Protective 15
1.8 Past Developments in Transformer Protective Schemes 15
1.8.1 Adaptive Digital Differential Protection for Transformer 15
1.8.2 DFT, FFT and Other Filtration Based Transformer Protective Schemes 17
1.8.3 Sequence Component-Based Transformer Protection Schemes 19
1.8.4 Artificial Intelligence (AI) Based Transformer Protection Schemes 20

1.8.5 Wavelet Transforms (WT) Based Transformer Protection Techniques . . . 22
1.8.6 Classifier Technique Based Transformer Protection Schemes . . . 23
1.8.7 All Other Methodology Used for Transformer Protection . . . 24
1.9 Combined Filtration and Classification Scheme for Transformer Protection . . . 25
1.10 Summary . . . 26
References . . . 26

2 CT Saturation Detection and Compensation Algorithm . . . 33
2.1 Proposed Method for CT Saturation Detection . . . 34
2.1.1 Proposed Algorithm . . . 34
2.1.2 Condition for CT Saturation Detection . . . 36
2.2 Proposed Saturation Detection Flowchart . . . 36
2.3 System Study . . . 38
2.4 Simulation Results and Discussion . . . 38
2.4.1 Effect of DC Component and Secondary Burden on CT Saturation . . . 39
2.4.2 Effect of Remanent Flux on CT Saturation . . . 40
2.4.3 Effect of Noise Superimposed in Secondary Current . . . 41
2.4.4 Effect of Types of Fault and Fault Inception Angle (FIA) . . . 42
2.5 Proposed Compensating Algorithm . . . 42
2.6 Practical Validation of the Proposed Algorithm . . . 44
2.6.1 Hardware Setup . . . 44
2.6.2 Results of Prototype . . . 44
2.7 Comparison of the Proposed Algorithm with Existing Scheme . . . 46
2.8 Summary . . . 47
2.9 Published Article Based on This Work . . . 48
Appendix . . . 48
References . . . 48

3 Phasor Angle Based Differential Protection of Power Transformer . . . 51
3.1 Literature Review . . . 51
3.2 A Proposed Transformer Protection Technique . . . 53
3.2.1 Problem Description and Solution . . . 53
3.2.2 Proposed Algorithm . . . 54
3.2.3 System Modeling . . . 59
3.3 Simulation Results with Discussion . . . 60
3.3.1 Inrush Condition . . . 60
3.3.2 Internal Fault in Transformer . . . 62

3.3.3 High Resistance Internal Fault 62
3.3.4 Internal Fault with Heavy CT Saturation 63
3.3.5 External Fault 65
3.3.6 External Fault with Heavy CT Saturation 65
3.4 Experimental Test Setup 68
3.4.1 Laboratory Prototype 68
3.5 Prototype Result Analysis 71
3.5.1 Inrush 71
3.5.2 Internal Fault 71
3.5.3 External Fault 72
3.5.4 External Fault with Deep CT Saturation 73
3.6 Novelty Projected in This Research Work 77
3.7 Summary 78
3.8 Published Article Based on This Work 79
Appendix 79
References 80

4 Adaptive Digital Differential Protection of Power Transformer 83
4.1 Literature Studied on Transformer Protection 83
4.2 Problem Discussion and Definitions 85
4.3 System Modeling 86
4.4 Proposed Adaptive Relaying Scheme 87
4.4.1 Third (3rd) Derivative-Based Technique 88
4.5 Result Discussion 91
4.5.1 Magnetizing Inrush Condition 91
4.5.2 Internal Fault on Transformer Winding 91
4.5.3 Transformer Internal Fault with CT Saturation 92
4.5.4 External Fault Condition 93
4.5.5 External Fault Condition with CT Saturation 93
4.6 Comparison of the Studied Results with Traditional Solution . . . 97
4.7 Hardware Implementation in Laboratory 98
4.7.1 Internal Fault Conditions 99
4.7.2 External Fault and Overload Condition 99
4.7.3 External Fault with Light, Medium and Heavy CT Saturation Conditions 101
4.7.4 Three Phase Transformer Hardware Results with Adaptive Shifting Characteristic Under CT Saturation Conditions 102
4.8 Novelty Introduced by the Proposed Scheme 104
4.9 Summary 104
4.10 Published Article Based on This Work 105
Appendix 105
References 105

5 Relevance Vector Machine Based Transformer Protection 107
5.1 Literature Studied for the Idea Generation 107
5.2 System Modeling and Data Generation 109
5.3 Proposed Transformer Fault Classification Methodology 113
5.3.1 RVM Classifier Model 113
5.3.2 SVM Learning Model 116
5.4 Proposed RVM Based Algorithm 117
5.5 Result Analysis and Discussion 119
5.6 Hardware Setup and Test Results 122
5.7 Advantages of the Proposed RVM Based Scheme 125
5.8 Summary 128
5.9 Published Article Based on This Work 129
Appendix 129
References 130

6 HE-ELM Technique Based Transformer Protection 133
6.1 Documentation of Comprehensive Review 133
6.2 System Modeling, Data Generation and Simulation 136
6.3 Existing and Proposed Techniques for Transformer Protection 139
6.3.1 PNN Learning Model 140
6.3.2 SVM Learning Model 141
6.3.3 ELM Learning Model 143
6.4 Proposed HE-ELM Learning Model 144
6.4.1 Feature Extraction Using Wavelet Transform 145
6.5 Proposed Fault Classification Algorithm 146
6.5.1 Parameter Tuning 146
6.6 Result Analysis and Discussion 147
6.6.1 Justification for Selection of the Size of Training Data Set in the Proposed Scheme 147
6.6.2 Classification Accuracy for Various Test Cases 148
6.7 Comparison of Proposed Techniques with Existing ELM, SVM and PNN Based Scheme 149
6.8 Hardware Setup and Test Results 156
6.9 Additional Tested DSO Results 160
6.10 Benefits of the Proposed Scheme 165
6.11 Summary 170
6.12 Published Article Based on This Work 170
References 170

7 Real-Time Monitoring and Adaptive Protection of Power Transformer 173
7.1 Literature Reviewed 173
7.2 Proposed Technique 175
7.2.1 Condition Monitoring of Transformer 175
7.3 Transformer Protection Approach 176
7.4 Experimental Test Setup and Result Discussion 180
7.4.1 Inrush Condition 181
7.4.2 Internal Fault 182
7.4.3 External Fault or Normal Condition 183
7.5 Monitoring of Other Transformer Conditions 185
7.6 Summary 188
7.7 Published Article Based on This Work 188
References 188

Conclusion 191

About the Authors

Dr. Dharmesh Patel is Assistant Professor in the Department of Electrical Engineering, Government Engineering College, Bharuch, Gujarat, India. He received a B.E. degree from Hemchandracharya North Gujarat University, Patan, Gujarat, in 1999, a master's degree in power system from the Sardar Patel University, Vallabh Vidyanagar, Anand, India, in 2002 and Ph.D. degree from Sardar Vallabhbhai National Institute of Technology, Surat, India, in 2019. His field of research is power transformer protection.

Dr. Nilesh Chothani is Associate Professor in the Department of Electrical Engineering at Adani Institute of Infrastructure Engineering, Ahmedabad, Gujarat, India. He received B.E. degree from Saurashtra University, Rajkot, Gujarat, in 2001. He received his master's degree in power system and the Ph.D. degree in electric engineering from the Sardar Patel University, Vallabh Vidyanagar, Gujarat, India, in 2004 and 2013, respectively. He has more than two decades of teaching experience.

He has published several papers in reputed international journals and conferences. Three of his research papers are awarded with work of excellence in *IEEE* conference. His areas of interest include digital protection, power system modelling and simulation, and artificial intelligence techniques. He has developed the state-of-the-art power system protection laboratory including real-time operation of digital/numerical relaying scheme. He also received a research grant funded by Science and Engineering Research Board (SERB), DST, New Delhi, Government of India.

Abbreviations and Symbols

Abbreviations

87 R	Differential relay
AAF	Anti-aliasing filter
ACF	Autocorrelation function
ADC	Analog-to-digital converter
AI	Artificial intelligence
APDP	Adaptive power differential protection
ATP	Alternative Transient Program
BC	Bayesian classifier
BFCL	Bridge-type fault current limiter
CBs	Circuit breakers
CRGO	Cold-rolled grain-oriented
CT/PT	Current transformer/potential transformer
CT_P and CT_S	Current transformer for primary and secondary of power transformer
DCMP	Differential current measuring principle
DOCC	DC offset current compensation
DSC	Digital signal controller
DSP	Digital signal processing
DWT	Discrete wavelet transform
E/F	Earth fault
EDP	Electromagnetic differential protection
EMTP	Electromagnetic transient programming
EWT	Empirical wavelet transform
FFBP	Feed-forward back propagation
FFT	Fast Fourier transform
FIA	Fault inception angle
FRIC	Fault-related incremental current
GA	Genetic algorithm

GIC	Geomagnetically induced current
GNC	Genetic neural computing
GT	Generator transformer
HA	Harmonic analysis
HE-ELM	Hierarchical ensemble extreme learning machine
HPF	High-pass filter
HRIF	High resistance internal fault
HRU	Harmonic restraint unit
HST	Hyperbolic S-transform
ILC	Improved lumped circuit
IMF	Intrinsic mode function
ISF	Instrument security factor
IT	Instantaneous trip
JA	Jiles–Atherton
KPV	Knee point voltage
LES	Last estimation square
LSSVM	Least square support vector machine
MATLAB	Matrix laboratory
MFCDFT	Modified full-cycle discrete Fourier transform
MI	Magnetizing inrush
MM	Mathematical morphology
MSNN	Master–slave neural network
MUB	Magnetic unbalance
NN	Neural network
O/C	Over-current
OLTC	On-load tap changer
OPNN	Optimal probabilistic neural network
OSHP	Optimal Separating Hyper-Planes
OTI/WTI	Oil temperature indicator/winding temperature indicator
PCA	Principal component analysis
PHA	Power and harmonic analyser
PNN	Probabilistic neural network
PSCAD	Power System Computer-Aided Design
PSO	Practical swarm optimization
PST	Phase-shifting transformer
RBF	Radial bias function
REF	Restricted earth fault
RTDS	Real-time digital simulator
RVM	Relevance vector machine
RVs	Relevance vectors
SAW	Symmetry assessment window
SC	Signal conditioning
SNR	Signal-to-noise ratio
SPR	Sudden pressure relay
STFT	Short-time Fourier transform

SVM	Support vector machine
SVs	Support vectors
SWDFT	Shorter-window discrete Fourier transform
TF	Transfer function
TL	Transmission line
TP/TN	True positive/true negative
TT	Time–time transform
TTF	Turn-to-turn fault
WPT	Wavelet packet transform
WTSE	Wavelet transform spectral energy
YY	Star–star connection of transformer

Symbols

50/51 R	Instantaneous and timed relay units (over-current)
64-	Earth fault relay unit
87	Differential relay
I_{rated}	Rated current
$I_{diff.}$ *or* I_d	Differential current
I_{bias} *or* I_b	Bias current
I_{d0} & I_{r0}	Basic differential and restraining current setting
$f(t)$	Sinusoidal current signal
fs	Sampling frequency
ΔT	Step of algorithm (period/time)(sampling time)
R_L	Load resistance
R_{ct}	Resistance at the CT secondary
K_s	Saturation factor
V_x	Saturation voltage
V_o	Output voltage
L_m	Magnetization inductance
I_m	Magnetizing current
K^{th}	Sampling signal
$F_{r(k)}$	Real part
$F_{i(k)}$	Imaginary part
A_s	Threshold for relay setting
θ	Phase angle
$\delta_{1(n)}$	First differential of equation
$\delta_{2(n)}$	Second differential of equation
$\delta_{3(n)}$	Third differential of equation
H	Sampling interval
X_s	Degree of saturation
K_1	Slope of relay
ΔM_s	Relative slope step

I_1 *or* I_p	Primary current
I_2 *or* I_s	Secondary current
V_1	Primary voltage
V_2	Secondary voltage
$F_{int.}$	Internal fault
$P_{(avg.)}$	Average power
$P_{(Reactive)}$	Reactive power
$P_{(Active)}$	Active power
Δ	Arctan of second-order derivative of differential current
Θ_{avg}	Average of arctan Δ
$[m_i, n_i]$	Different time interval over Θ_{avg} Estimated
$F_{ext.}$	External fault
$F_{int.}$	Internal fault
θ_d	Phasor angle difference b/w primary and secondary current
α	(Decaying coefficient) or (voltage angle)
ω	Angular velocity
$\emptyset$	Switching instant or fault inception angle (FIA)
C	Initial values of exponential component
$I_{2ndHarmonic}$	2nd harmonic current component
I_{Fund}	Fundamental component of current
I_{sat}	Saturated current
$I_{non\text{-}sat}$	Non-saturated current
Xi	Input data
W_i	Synaptic weights
$\emptyset_k$	Activation function
$h_{[n]}$	HPF coefficient
$f_{[n]}$	Discrete input signal
N	Circular window length
a	Operator
$\emptyset_r$	Residual flux
$\emptyset_m$	Maximum flux
abc	Three-phase stationary coordinate system
$d\text{-}q$	Two-phase rotating coordinate
$f_{RVM}(x)$	RVM classifier function
Δ	Second-order derivative of differential current
$[mi, ni]$	The different time interval over the average value
θ_{avg}	Average of the calculated angle
F_L	Fault location
R_f	Fault resistance
δ	Load angle
ρ	Sigmoid logistic function
ω	Weight vector
$P(\tau/\delta)$	Likelihood factor
τ	Target vector

χ_i	Hyperparameter
J	Objective function
$f_{SVM}(\delta)$	SVM classifier
V_m	Maximum value of applied voltage
$N_{pri.=}$	Primary turns
H	Current harmonic number
P_{EC}	Eddy current losses
X_h	Harmonic current
K_h	Harmonic constant
$\emptyset_{dc}^{max}$	DC maximum flux
y, r *and* d	Kernel parameters
$K\ (x_i,\ x_j)$	Kernel function
σ	Standard deviation
x_{ij}	jth training vector for class k_i
M_i	Number of training pattern in class k_i
ζ_i	Slack variables
F_f	Fitness function
Wf_i	Weight factor
S_i	Score of parameters
S_{max}	Maximum score of parameter
P_d	Differential power
P_r	Restraining power

List of Figures

Fig. 1.1 Classification of transformer fault and relevant protection 2
Fig. 1.2 Over current protection of transformer winding 5
Fig. 1.3 Overcurrent relay with harmonic restraint unit 6
Fig. 1.4 Restricted earth fault protections.......................... 7
Fig. 1.5 Circulating current differential protection 8
Fig. 1.6 Biased differential protection of transformer................. 8
Fig. 1.7 Typical dual slope percentage biased characteristics.......... 9
Fig. 1.8 Effect of magnetizing inrush............................. 10
Fig. 1.9 Winding and oil temperature indicator with alarm unit........ 13
Fig. 1.10 Buchholz relay and its magnified view..................... 14
Fig. 1.11 Overall arrangements of protective schemes for typical grid transformer 16
Fig. 1.12 DFT/FFT based algorithm................................ 18
Fig. 1.13 Sequence component-based algorithm...................... 19
Fig. 1.14 ANN-based algorithm.................................... 20
Fig. 1.15 WT based algorithm 22
Fig. 1.16 SVM based algorithm.................................... 23
Fig. 2.1 Algorithm of CT saturation detection 37
Fig. 2.2 Single line diagram of the power system model.............. 38
Fig. 2.3 Waveform of CT currents and value of D_n and T_h, **a**, **b** without CT saturation, **c**, **d** with CT saturation 39
Fig. 2.4 Waveform of CT currents and value of D_n and T_h under CT saturation condition, **a**, **b** $R_b = 3\ \Omega$ and **c**, **d** $R_b = 5\ \Omega$ 40
Fig. 2.5 Waveform of CT currents and value of D_n and T_h during **a**, **b** 0% remanence flux and **c**, **d** 90% remanence flux 40
Fig. 2.6 Waveform of CT primary and secondary current **a** and value of D_n and T_h, **b** during SNR = 40 db contained by CT secondary signals.. 41
Fig. 2.7 Waveform of CT currents and value of D_n and T_h during **a**, **b** FIA $\theta = 45°$ and $R_b = 3\ \Omega$ and **c**, **d** FIA $\theta = 135°$ and $R_b = 5\ \Omega$, respectively.. 42

Fig. 2.8 **a** CT currents and estimated current magnitude by MDFT filter, **b** compensated current magnitude, and **c** compensated phase angle of the CT for the current signal of Fig. 4c 43
Fig. 2.9 Hardware setup of laboratory test bench.................. 45
Fig. 2.10 **a** CT secondary current captured by DSO and **b** values of D_n and T_h for the said condition 45
Fig. 2.11 **a** CT currents, **b** value of del_2 and T_{h1} during second difference, **c** value of del_3 and T_{h2} during third difference, **d** value of D_n and T_h of the proposed algorithm 46
Fig. 2.12 **a** CT currents, **b** output of wavelet technique and **c** value of D_n 47
Fig. 3.1 Phasors of primary and secondary current during **a** Internal and **b** External fault 54
Fig. 3.2 Proposed algorithms for transformer protection............. 55
Fig. 3.3 Circuit diagram of power system 59
Fig. 3.4 Inrush condition **a** Primary and secondary current of transformer, **b** Arc tan of Δ and **c** average of angle (θ_{avg})..... 61
Fig. 3.5 Inrush followed by internal fault **a** Primary and secondary current of transformer, **b** Arc tan of Δ and **c** average Arc tan of Δ ($\theta_{avg.}$) 62
Fig. 3.6 Internal fault **a** Primary and secondary current, **b** differential and restraining current, **c** phasor angle of currents 63
Fig. 3.7 High resistances internal fault **a** Primary and secondary current, **b** differential and restraining current, **c** phasor angle of currents.................................. 64
Fig. 3.8 Heavy CT saturation in internal fault **a** Primary and secondary current, **b** differential and restraining current, **c** phasor angle of currents.................................. 66
Fig. 3.9 External fault **a** Primary and secondary current, **b** differential and restraining current, **c** phasor angle of currents 67
Fig. 3.10 Heavy CT saturation in external fault **a** Primary and secondary current, **b** differential and restraining current, **c** phasor angle of currents.................................. 68
Fig. 3.11 Prototype model developed in laboratory for transformer protection 70
Fig. 3.12 Magnetising inrush 71
Fig. 3.13 Internal fault **a** Value of primary and secondary current magnitude and phase angle, **b** phasors of primary and secondary current, **c** waveform of primary and secondary current during fault.............................. 72
Fig. 3.14 Internal fault **a** Waveform of primary and secondary current, **b** phasor of primary and secondary current during internal fault 73

Fig. 3.15 External fault **a** Value of primary and secondary current magnitude and phase angle, **b** phasors of primary and secondary current, **c** waveform of primary and secondary current during fault 74
Fig. 3.16 External fault with CT saturation **a** Value of primary and secondary current magnitude and phase angle, **b** phasors of primary and secondary current, **c** waveform of primary and secondary current during CT saturation 74
Fig. 3.17 CT saturation under external fault **a** Waveform of primary and secondary current, **b** phasor of primary and secondary current during fault 75
Fig. 4.1 Two-stage biased differential relay characteristics 85
Fig. 4.2 Line diagram for testing 86
Fig. 4.3 Proposed fault zone identification algorithm 90
Fig. 4.4 Magnetizing inrush condition, **a** primary current and secondary current, **b** fundamental and second harmonic components 92
Fig. 4.5 Internal fault, **a** primary versus secondary current, **b** magnitude of differential and restraining current, **c** I_{diff}/I_{bias} trajectory without fault resistance, **d** I_{diff}/I_{bias} trajectory with 10 Ω fault resistance 93
Fig. 4.6 Internal fault with CT saturation, **a** transformer primary and secondary current, **b** magnitude of differential and restraining current, **c**, **d** I_d/I_{bias} trajectory with medium and heavy CT saturation respectively 95
Fig. 4.7 External fault, **a** primary versus secondary current, **b** magnitude of differential and restraining current, **c** I_{diff}/I_{bias} trajectory 96
Fig. 4.8 I_{diff}/I_{bias} trajectory under various condition, **a** mild CT saturation, **b** medium CT saturation, **c** current during heavy CT saturation, **d** trajectory during heavy CT saturation 96
Fig. 4.9 External fault with heavy CT saturation, **a** transformer primary and secondary current, **b** magnitude of differential and restraining current, **c** I_d/I_{bias} trajectory with existing scheme [24] and proposed scheme 98
Fig. 4.10 **a**$_1$, **b**$_1$, **c**$_1$ Primary and secondary current waveform during internal fault and **a**$_2$, **b**$_2$, **c**$_2$ I_d/I_{bias} trajectory for internal fault with zero resistance, CT saturation under internal fault, high resistance internal fault 100
Fig. 4.11 **a**$_1$, **b**$_1$ Recorded primary and secondary current waveform and **a**$_2$, **b**$_2$ I_d/I_{bias} trajectory for external fault and overloading condition 100
Fig. 4.12 **a**$_1$, **b**$_1$, **c**$_1$ External fault current waveform during low, medium and heavy CT saturation and **a**$_2$, **b**$_2$, **c**$_2$ I_d/I_{bias} trajectory for low, medium and heavy CT saturation under external fault conditions 101

Fig. 4.13 Three phase hardware setup L-L fault (with one CT saturated) DSO results and shifting of adaptive percentage biased characteristics . . . 103
Fig. 5.1 Single line diagram for Indian power system . . . 109
Fig. 5.2 Types of inrush . . . 118
Fig. 5.3 Proposed RVM based fault classification algorithm . . . 122
Fig. 5.4 Primary and secondary current waveform under **a** inrush condition **b** internal fault **c** external fault and **d** CT saturation condition . . . 123
Fig. 5.5 Current signals during different fault/inrush conditions . . . 123
Fig. 5.6 Hardware setup in the laboratory for transformer fault analysis . . . 124
Fig. 5.7 Circuit diagram and control circuit of hardware setup . . . 126
Fig. 5.8 Primary and secondary current waveform for **a** inrush **b** internal fault **c** external fault **d** external fault with CT saturation . . . 128
Fig. 6.1 Single line diagram of the Indian power system . . . 136
Fig. 6.2 Structure of PNN . . . 141
Fig. 6.3 Proposed HE-ELM technique based algorithm . . . 147
Fig. 6.4 Graph of training data versus percentage accuracy . . . 148
Fig. 6.5 Hardware prototype in laboratory **a** front view, **b** rear view of the panel . . . 155
Fig. 6.6 Three phase diagram (with control diagram) for hardware set up to create fault and abnormalities on considered power transformer . . . 157
Fig. 6.7 Detailed view of laboratory setup . . . 158
Fig. 6.8 Transformer primary and secondary side current waveform for case **a** Inrush **b** internal fault (L-G) **c** internal fault (LLg) **d** external fault (LLL) **e** external fault (L-G) with CT saturation . . . 161
Fig. 6.9 Transformer primary side current waveforms for inrush condition . . . 162
Fig. 6.10 Transformer primary side current waveforms for internal (L-G) fault condition . . . 162
Fig. 6.11 Transformer primary side current waveforms for internal (L-G) fault condition with low fault resistance . . . 163
Fig. 6.12 Transformer primary side current waveforms for internal fault condition (L-G fault with slight decaying DC component) . . . 163
Fig. 6.13 Transformer primary side current waveforms for internal (LL-G) fault condition . . . 164
Fig. 6.14 Transformer primary side current waveforms for internal (LL) fault condition . . . 164
Fig. 6.15 Transformer primary side current waveforms for internal (LLL) fault condition on lower tapping . . . 165

Fig. 6.16 Transformer primary side current waveforms for internal (LLL) fault condition on higher tapping 166
Fig. 6.17 Transformer primary side current waveforms for external (L-G) fault condition. 166
Fig. 6.18 Transformer secondary side current waveforms for external (L-G) fault condition. 167
Fig. 7.1 Generalized schematic diagram for transformer monitoring and protection. 175
Fig. 7.2 Proposed Adaptive Power Differential Protection (APDP) scheme 177
Fig. 7.3 Differential power versus restraining power characteristic. 178
Fig. 7.4 Proposed adaptive PDP based algorithm. 179
Fig. 7.5 Developed laboratory setup. 181
Fig. 7.6 **a** Voltage waveform during inrush. **b** RMS value of voltages during inrush. **c** Voltage waveform during fault. **d** RMS value of voltages during fault. 182
Fig. 7.7 Inrush condition. **a** Three phase inrush currents waveform. **b** Per phase harmonic during inrush. **c** Spectrum analysis during inrush 183
Fig. 7.8 Internal fault conditions 184
Fig. 7.9 External condition. **a** Current waveform. **b** Voltage waveform. **c** RMS value of voltages 185
Fig. 7.10 Differential versus restraining power characteristic during external fault condition, **a** without CT saturation. **b** with CT saturation 186
Fig. 7.11 **a** Parameter variation versus fitness function, **b** loading versus efficiency, **c** loading versus temperature and **d** loading versus losses 186

List of Tables

Table 3.1 Various fault and system parameter values considered 60
Table 3.2 Current and phasor comparison of primary and secondary current in internal fault 65
Table 3.3 Test current and phasor comparison of primary and secondary current in external fault 69
Table 3.4 Test conditions validation through prototype model 76
Table 4.1 Performance of the proposed algorithm during different types of internal faults 94
Table 5.1 Training and testing data considered for various internal faults 110
Table 5.2 Training and testing data considered for various external faults 111
Table 5.3 Training and testing data generated for various inrush conditions 112
Table 5.4 Total training and testing data collection for various conditions 113
Table 5.5 Empty feature vector for training datasets 114
Table 5.6 Classification accuracy for different fault cases 119
Table 5.7 Fault type wise classification accuracy 120
Table 5.8 Comparisons of the proposed RVM Scheme with SVM and PNN scheme 121
Table 5.9 Fault data generation using hardware setup 127
Table 6.1 Training and testing data generated through various internal fault conditions 137
Table 6.2 Training and testing data for various external faults 138
Table 6.3 Training and testing data generated for various inrush conditions 139
Table 6.4 Total training and testing data collection for various conditions 139
Table 6.5 Classification accuracy of the proposed scheme with varying training and testing data size 148

Table 6.6 Classification accuracy of the proposed HE-ELM scheme for different fault cases 150
Table 6.7 Fault category wise classification accuracy using HE-ELM 150
Table 6.8 Cross-validation of the proposed scheme for different training and testing data 151
Table 6.9 Comparisons of the proposed HE-ELM scheme with SVM and PNN scheme 155
Table 6.10 Fault data generation through hardware setup 160
Table 7.1 Parameters and respective weight factors for the defined fitness function 176
Table 7.2 Fitness function (F_f) for change in transformer parameter 187

Chapter 1
Introduction to Power Transformer Protection

1.1 Introduction

India is a leading country and its economy grows day by day by attracting foreign direct investments (FDI). Production, manufacturing, industrial, software, and all other business depends on reliable electricity supply with a slogan of "without power-no business", shows the importance of reliable power supply. The protection of the power system is a very sensitive and burning issue due to huge expansion, complexity, and deregulation. Future power reliability with growth in power generation, expansion, and improvement as per nation demand is the main challenge for India. For transferring power in a grid, the transformer work as the heart of the power system. Having critical importance of power transformer, unwanted failure generates critical issues not only for industrial & other customers but also affects the national economy, social and political concern. Power transformer failure analysis of Maharashtra state (India) [1] gives the main exposure to investigate the causes of failure and focus on various transformer protective schemes. Also, Binder [2] involved transformer failure analysis for investigators and researchers with trends and scope of transformer failure.

Reliability and the fast protective scheme is the main requirement due to an important role of a power transformer. The non-linear core characteristic of a transformer is one of the main issues in power and current transformers. It is very difficult to protect the system against the core saturation. Nonlinearity in power transformer generates magnetizing inrush and in current transformer secondary current gets saturation so accuracy is reduced to measure actual quantity. The peak value of the current is not only generated due to overload or under fault conditions but also due to harmonics and resonant conditions generated by core nonlinearity.

Due to an issue of sensitivity in the power system, complete transformer protection is a very strong issue in the HVAC system. Normally for 132 kV and above grid system protective schemes needs high-speed fault clearance for stability point of view and reduce damage due to fault [3]. A target of this book chapter is providing foundation knowledge regarding transformer protection and collective information of various

D. Patel and N. Chothani, *Digital Protective Schemes for Power Transformer*, Power Systems, https://doi.org/10.1007/978-981-15-6763-6_1

schemes of transformer protection utilized by past research scholars with advantages and limitations with proposed techniques. Finally, a motto of research is providing complete and accurate transformer protection with the least computational burden, trivial complexity, and minimum time of operation.

Nowadays, an electrical power system is becoming more complex due to an increase in transmission line length and associated equipment to meet additional power requirements. Due to the complexity of the power system and financial constraints, schemes of the protective relays are also facing many problems. A transformer is the heart of a substation. It is used to convey the power from one circuit to another circuit without changing frequency. Approximately 10% of faults occur on the transformer which is described in fault statistics of a power system [4]. For providing protection to a power transformer, all details are required like kVA rating, voltage ratio, windings information, percentage reactance, resistance, earthing resistance, indoor or outdoor, dry or oil-filled, with or without conservator, also length and cross-section of connecting leads between CT's and relay panel, fault level at power transformer terminals, network diagram showing the position of the transformer [5] is required. Also, IEEE guidelines [6] are provided basic information for transformer protection by the IEEE power engineer society.

Generally, transformer protection is categorized on the basis of operating voltage and volt-ampere range of a transformer. For providing protection, the power transformer is generally categorized in three parts (1) Small power transformers which have a rating up to 500 kVA (2) Medium power transformers which have a rating more than 500 kVA to 5 MVA (3) Large power transformers have a rating greater than 5 MVA.

Transformer failure statistics show that maximum transformer failure is due to winding failure and tap changer failure [7]. Transformer protection is categorized in two ways (1) Electrical (2) Non-electrical as shown in Fig. 1.1.

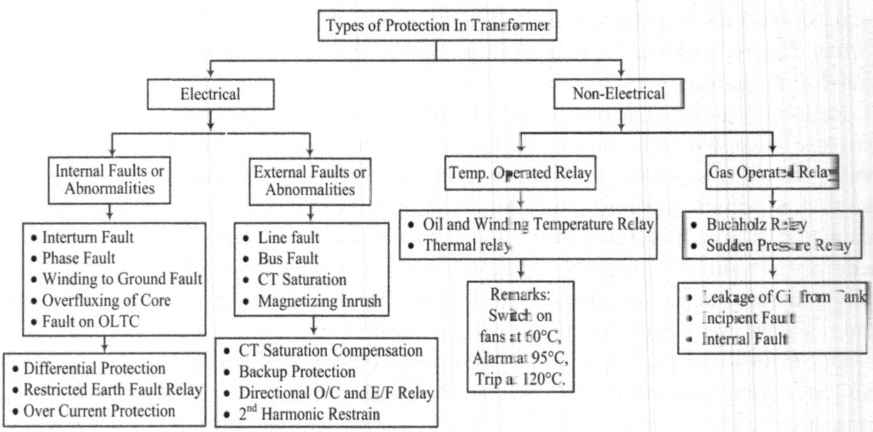

Fig. 1.1 Classification of transformer fault and relevant protection

1.2 Types of Faults and Abnormalities

Various types of faults and abnormal conditions arise on transformers [8] which described in the subsequent section.

1.2.1 Internal Fault

Internal faults are subdivided into two groups, (a) major faults or active faults and (b) minor faults. Depending on the severity of the internal fault, there is a risk of fire and damage to winding and in the worst-case entire body of a transformer. This is due to the electromagnetic and mechanical force developed in the winding and oil of the transformer. This phenomenon leads to the loss of costly equipment and loss of power supply to the connected line for a long time.

Group (a) major faults

Major faults produce quick damage to a transformer and affect the entire power system network. Generally, these types of faults are detected by the unbalancing quantity of voltage and current. Such faults are a phase to ground, phase to phase, double phase to the ground on a high voltage and low voltage bushings, phase to earth or phase to phase fault on a high voltage and low voltage winding, and short-circuit between a high voltage and low voltage turns. Ground faults or a tertiary winding or turns short circuit in a tertiary winding, core faults, tank faults.

Group (b) minor faults

Minor faults or incipient faults are causing slow damage in the equipment or developed by damaged on equipment. This situation cannot be detected with the help of unbalance parameters of voltage and current. Generally, they include bad or poor electrical connection or faults on a core, which causes limited arcing under the oil. Failure of coolant generates high temperatures even under load conditions. Due to low oil content or if oil flow is clogged, which cause local hot- spot on winding. If percentage impedance is different in parallel connection of transformers than unequal load sharing may cause overheating. Weak insulation may cause leakage between winding and core and may result in a severe fault.

As discussed earlier due to the severity of fault in a group (a), faulted equipment or part must be isolated as fast as possible within minimum disturbances. The faults of a group (b) are not very serious in their initial stage, but they may develop major faults later on if persistent for a long time. Hence, it must be cleared within a short time to preserve system reliability.

Causes of Internal Fault

Transformer failures are normally initiated as follow [9]:

Winding breakdown: Reasons for failures of the winding are insulation deterioration, manufacturing defects, overheating, voltage surges, mechanical stresses, and vibrations. Terminal board and on-load tap changer failure due to improper assembly or improper design, damage in transportation, or high vibration. Bushing failure is due to aging, cracking, animal hunting, contamination, and vandalism. Load tap changer fails because of mechanical malfunctioning, a problem in contacts, vibrations, insulating liquid contamination, improper assembly, and high stress. Miscellaneous failures due to CT bushing failure, core insulation failure, oil leakage due to tank damage or poor welding, the presence of foreign material in the tank, or shipping damages.

1.3 External Fault or Abnormalities

External faults mean faults occurring outside the transformer protection zone and other abnormalities that are subdivided into an overload condition, overvoltage, under frequency, and magnetizing inrush [7]. Though the abnormalities are not faults in a transformer they result in overheating, insulation damage, increased oil pressure which leads to generating a situation of an explosion. An external fault causes CT saturation and malfunctioning of the protective schemes of a transformer. Even large external fault current causes large mechanical stress on transformer windings. Thus, the external short circuit should be detected and discriminated from an internal fault in transformer protection.

An overload condition is detected by thermal relays which give alarm so that this condition is attended by a supervisor. Overload condition causes overheat, reduces the lifespan of the equipment, and may cause permanent damage. One of the main causes of overheating is the unequal loading of the three-phase system on a transformer. Overvoltage is divided in the short term and long term transient condition and this transient overvoltage cause stress on end turn of the winding. Due to an emergency operating condition like a sudden loss in load, power frequency overvoltages occur, which creates over fluxing in the transformer (V/f). This increases stress on winding, rise in the iron loss and also increase in heating of the iron core. So insulation of lamination and winding may get damage during overvoltage.

Under frequency arises in the system due to major disturbances such as the imbalance between load and generation. Over fluxing relay is energized and provides a trip signal. Normally "Volts per hertz" (V/f) limit should not exceed 1.[illegible] per unit. Magnetizing inrush situations happen during the energization of the transformer under no-load condition. The magnitude of current during this condition depends on the switching instant and remnant flux sustained by the core of a transformer. Though the magnitude of inrush current is as high as fault current the transformer protective scheme should remain stable during this condition.

1.4 Various Protective Schemes Used in Power Transformers

In view of the stability and reliability of a power system with consideration of the cost and importance of a transformer, it is advisable to provide protective schemes on the transformer. Various protective schemes are described in subsequent sections.

1.4.1 Over Current Protection

Figure 1.2 shows the connections for an overcurrent (O/C) protection scheme for one of the transformer winding. Normally overcurrent protection is not preferred by the manufacturer of a transformer, but it is preferred as backup protection and often used as main protection in small transformers. In this scheme, an extremely inverse characteristics type overcurrent relay is preferred with an instantaneous unit for severe faults. Instantaneous protection is provided by the O/C relay at 400% and above-rated current. Three O/C units for phase fault and one earth fault (E/F) unit for a ground fault is used to protect small size transformers.

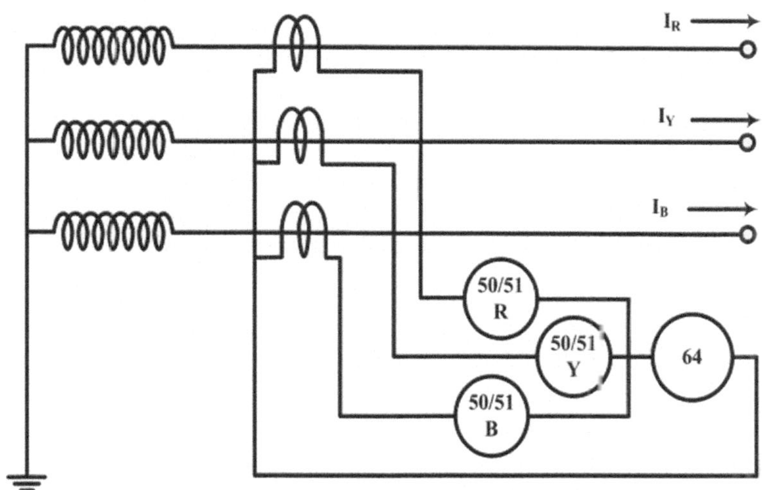

- **50/51- Instantaneous and Timed Relay Units**
- **64- Earth Fault Relay Unit**

Fig. 1.2 Over current protection of transformer winding

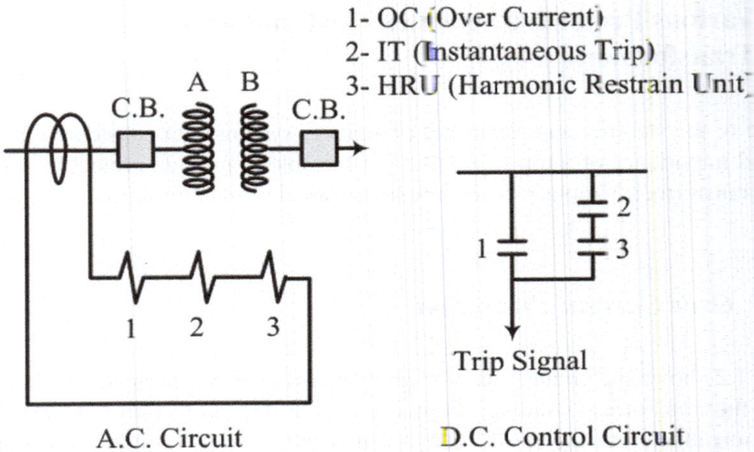

Fig. 1.3 Overcurrent relay with harmonic restraint unit

1.4.2 Overcurrent Protection with Harmonic Restraint Unit (HRU)

Figure 1.3 shows the overcurrent relay with the harmonic restraint unit. This scheme gives high-speed tripping when a transformer is energized during the fault. Per phase single harmonic restraint unit (HRU) with instantaneous trip (IT) elements are used to supplement the time delay overcurrent (O/C) relay [10].

Generally, this type of protection is used to avoid unnecessary tripping of the transformer during magnetizing inrush conditions by HRU and successfully operates during the transformer switched under a fault condition. This scheme is suitable for a small rating transformer where differential protection is not affordable.

1.4.3 Restricted Earth Fault (REF)

For a Y-connected medium rating transformer having a neutral grounded winding, Restricted Earth Fault (REF) protection is used as main protective schemes. This scheme provides protection to the internal ground fault of the Y-connected transformer winding. Connections for the REF scheme are shown in Fig. 1.4.

Even during a high magnitude external fault, if the proper CT ratio is selected, the relay remains in an inoperative condition [8].

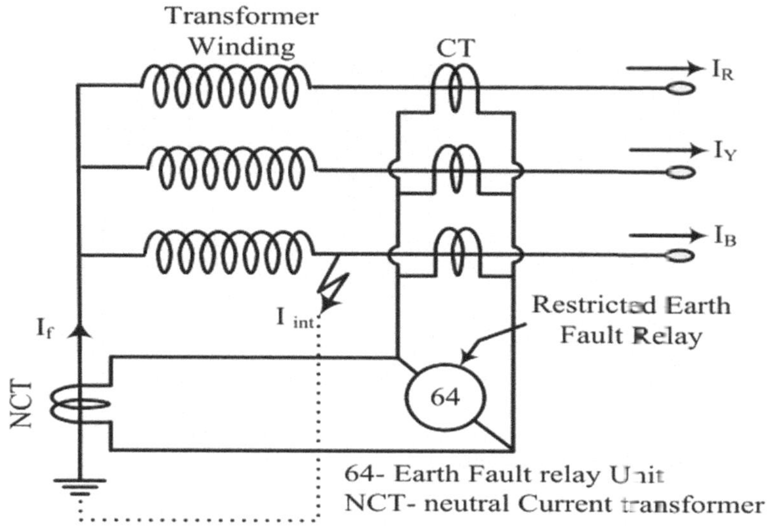

Fig. 1.4 Restricted earth fault protections

1.4.4 Differential Protection

Differential protection of a transformer is divided into (a) circulating current differential protection and (b) percentage biased differential protection.

(a) **Circulating Current Differential Protection**

The circulating current differential based relaying scheme of a power transformer is the simplest form of protection. Figure 1.5 shows the connection between CTs and relay for the circulating current differential scheme. Transformers having more than 10 MVA rating use differential protection scheme. However, these relays cannot be sensitive as the differential relays are used in generator and busbar protection. The phase-shifting in the star-delta transformer should be taken as main the factor otherwise protection scheme may mal-operate. Mismatch of CT ratios, different voltage ratings, magnetizing inrush currents, CT saturation is few other causes for maloperation of a differential scheme.

Transformer protection is also more complicated in multi-winding transformer banks, zigzag transformers, and transformer in-unit systems, phase angle regulators (PAR), voltage regulators, and 3-phase transformer banks composed of single-phase units.

(b) **Percentage Biased Differential Protection Schemes**

To avoid mal-operation of a simple differential protection scheme in star-delta connection of a transformer due to resistance variation of different lead length and CT

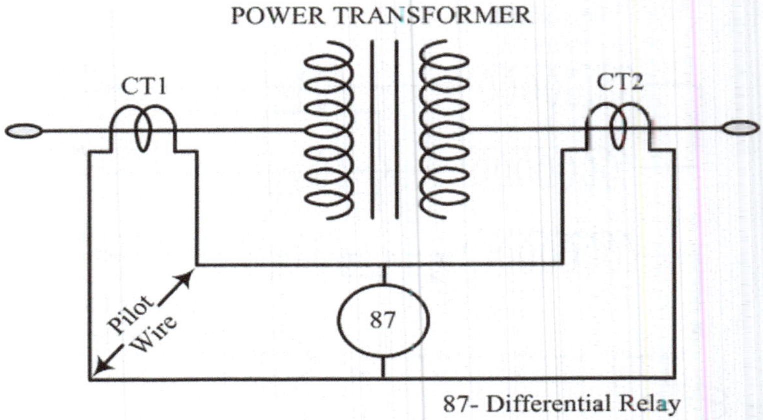

Fig. 1.5 Circulating current differential protection

secondary resistance and phase shifting of both winding, biased percent differential protection is applied as shown in Fig. 1.6.

Still, some problems are observed while applying for percentage bias differential protection in a transformer. These are listed as magnetizing inrush, CT saturation, high resistance internal and external fault condition, power swing conditions in the power system, effect of a harmonic. In the conventional differential protective scheme, a single slope characteristic is preferred for the medium-range transformer. Generally, a transformer differential relay is not sensitive with respect to other unit differential protection since many constraints are applicable, even restraining force is also higher due to some reason.

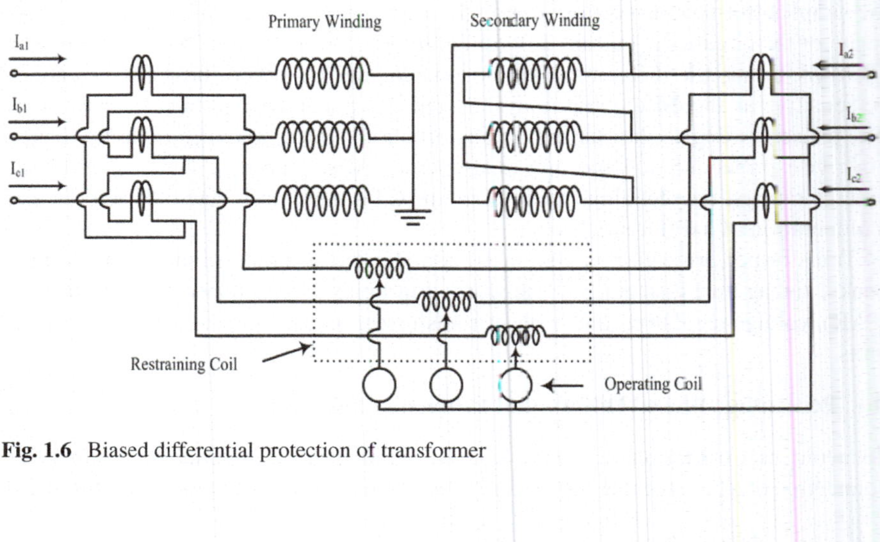

Fig. 1.6 Biased differential protection of transformer

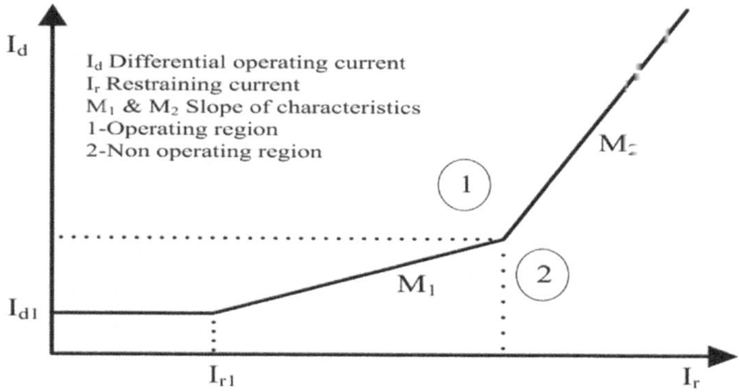

Fig. 1.7 Typical dual slope percentage biased characteristics

A digital differential protective scheme is subdivided into single slope and dual-slope characteristic as shown in Fig. 1.7. Normally, the slope of the characteristics M_1 lies between 0.4–0.7, and M_2 is 0.5–0.75 as described in Fig. 1.7 [11].

Digital differential technology improves reliability, increases dependability and security, provides self-checking facilities. Within moderate cost, they give high performance, even reduce a burden on CT and PT and also provide higher flexibilities with respect to conventional relays. Numerical relays offer very less burden to secondary of the CT so the performance of CT is improved during fault [12]. Proper selection of the ratio of the current transformer and Knee Point Voltage (KPV) will reduce exposure to the problems of CT saturation [13].

1.5 Burning Issues for Transformer Protection

During the protection of transformer Magnetizing Inrush, CT saturation conditions, over-fluxing conditions, and inter-turn fault detections are major burning issues in the real field of a power system.

1.5.1 Magnetizing Inrush Phenomenon

Transformer Magnetizing Inrush (MI) is a burning issue since the AC system developed. In 1944 Brownlee [14] and Blume [15] elaborated transformer magnetizing current broadly with effect on a power system. Again Holcomb [16] elaborated on the effect of MI on distribution transformer. The effect of MI on the transformer protection relaying scheme is elaborated by Van Warrington [8]. Recently, cold rolled grain oriented (CRGO) silicon steel is used as a core material with a saturated flux density of around 2.0 T. Even continuous improvement is going on for improving Volt-Amp per kg and power per kg characteristics [17]. At instant switching, voltage wave corresponds to flux density in a core. Residual magnetism (flux) also shares a key role in core saturation.

The peak value of flux in core [4] is

$$\varnothing = \varnothing_r + \varnothing_m \cos\theta + \varnothing_m \cos(\omega t + \theta) \tag{1.1}$$

So, the transformer flux is a function of the following factors:

Residual flux $\varnothing_r$, Maximum flux $\varnothing_m$, Switching instant angle θ, Core magnetic properties. From Eq. 1.1, we see that for $\theta = 0$ and $\varnothing_r = \varnothing_m$ the flux achieve an amplitude of $3\varnothing_m$ at $\omega t = \pi$ radians. To assure a flux demand of $3\varnothing_m$, the transformer primary draws a very huge magnetizing current with a peak non-sinusoidal waveform. The phenomenon of magnetic inrush is shown in Fig. 1.8.

Generally, three types of magnetizing inrush conditions are described such as initial, recovery, and sympathetic inrush as per transformer connection and its switching in the power system. Mostly 2nd harmonic component-based inrush detection techniques are utilized in past however, 2nd derivative of differential currents [18] are also utilized to discriminate inrush as abnormal conditions.

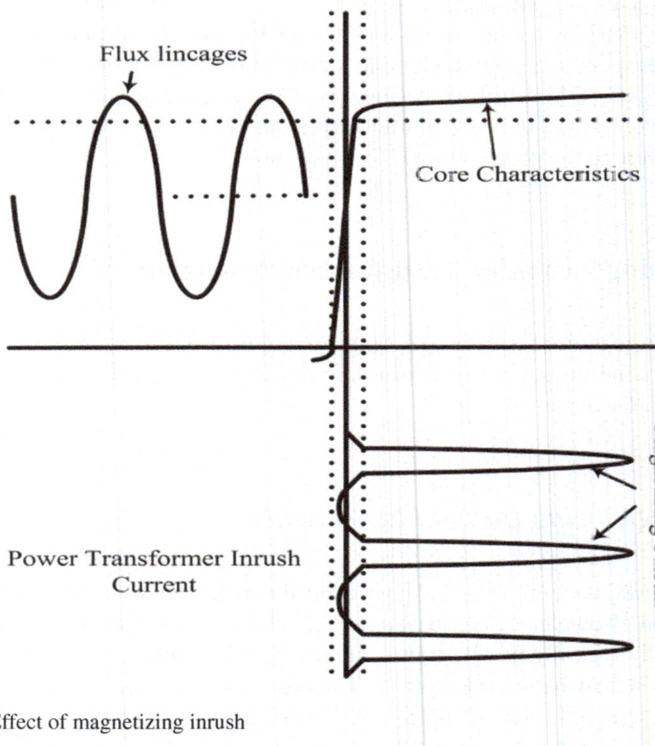

Fig. 1.8 Effect of magnetizing inrush

1.5.2 Current Transformer Saturation Conditions

Current Transformer (CT) saturation condition is a burning issue for unit type transformer protection. Effect of CT saturation is a major drawback in transformer differential protection, even detection of CT saturation and compensation techniques is also a complicated process. Magnetizing inrush and CT saturation generate major complication to provide reliable protective signals.

Currently available schemes of transformer protection face many types of adverse effects due to unfaithful Current Transformer (CT). CT Saturation, CT ratio Mismatch, Measuring Equipment Errors, Fault Inception Angle (FIA), higher burden, remnant flux, etc. are the major issue for protective CT. Without having a thorough knowledge of the relay, one cannot predict the performance of relay in non-sinusoidal current waveforms. Electromagnetic, static, and digital relays give special effects on CT saturation [19]. CT operations under the nonlinear region CTs are generated more complacency and its discriminations are also a major issue.

1.5.3 Over Fluxing Condition

The flux and the applied voltage in a transformer are related as per the following expression of EMF induced in a transformer.

$$V = 4.44 * \varnothing_m * f * N \tag{1.2}$$

where, V is the RMS value of the voltage, Φ_m maximum flux, f is the frequency, N is the number of turns in the winding. Thus, we can write the flux as

$$\varnothing_m = \frac{V}{4.44 * \varnothing_m * f * N} \tag{1.3}$$

The transformer core gains higher flux to tackle the overvoltage condition (keeping frequency constant). From a design point of view, power transformers work at the knee point of the magnetization curve at normal voltage. So, any rise in applied voltage and the subsequent rise in flux density drives the transformer into saturation region. This condition is described as over excitation during which the transformer draws too much magnetization current. A volt per hertz relay is used to detect an overfluxing situation by measuring the V/f ratio of a transformer. In interconnections of a power system, transformer over fluxing protection is implemented on both HV/LV sides.

1.5.4 Inter-turn Fault Protection

Due to inter-turn faults, heavy current flows inside the transformer through shorted turns. As seen from the transformer terminal, the measured current during inter-turn fault may be relatively the same on both sides [4]. Thus it is difficult to identify this situation using differential protection. Even, they can cause severe hot spots ensuing in deterioration of insulation and oil. Buchholz relay is used to detect an inter-turn fault by means of decomposition of oil due to heat. The consequent gas generated is used to sense the fault by purely nonelectrical means. A rate of rising of pressure relay provides the highest sensitivity [10] against inter-turn fault, which is covered in non-electrical protection.

1.6 Non-electrical Protection

Some fault in a transformer grows slowly, they can decompose oil and insulation and leads to major arcing faults. In order to protect the transform against minor fault and incipient faults, non-electrical protection is required. Types of non-electrical faults are explained in a subsequent section.

1.6.1 Thermal Relays

Usually, in a transformer, thermal protection is arranged to alarm about the panic of a circuit after the requisite time delay and in worst condition tripping. Thermo-couples or resistance temperature detectors are used to measure the oil and winding temperature. Actually, two types of indicators are provided on large transformers as shown in Fig. 1.9, (1) Oil temperature indicator (2) Winding temperature indicator (hot spot thermometer).

The oil temperature is measured directly by an RTD sensor kept in a transformer tank in touch with oil at the top of the transformer. The winding temperature is measured by inserting a small current transformer (CT) in series with the main winding of the transformer. The secondary of this CT and sensing bulb (RTD) measures the temperature proportional to the current flow through the windings. The output leads of the sensing bulb/RTD are connected to the oil temperature and winding temperature inductor (OTI/WTI) and alarm/protective unit. Figure 1.9 shows the connection between OTI and WTI with the transformer. When the temperature of any RTD crosses the threshold value, an alarm is actuated.

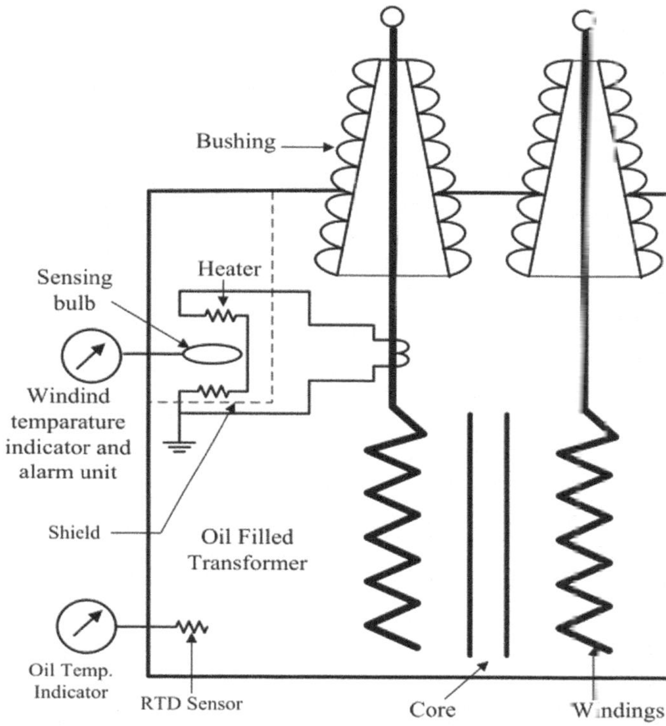

Fig. 1.9 Winding and oil temperature indicator with alarm unit

1.6.2 Buchholz Relay

Figure 1.10 shows a Buchholz relay connected between conservator tank and transformer, to detect gas produced in the transformer due to oil decomposition.

The conservator pipe must be placed with a slight inclination for reliable operation. As the gas gathers, the oil level falls and float F operates mercury switch with sounding an alarm. Small incipient and slowly rising faults can be identified by the Buchholz relay. The relay gives an alarm when the gases accumulated have reached a specific volume, which depends upon the transformer size. When a winding fault occurs, the arc produces gas at a speed over 50 $cm^3/kW/s$ which creates a surge in the oil.

This quickly moves the vane (V) and causes tripping through contacts close to the vane (Fig. 1.10) [8]. The angle of displacement of the mercury switch for making contact is about 15° plus the angle of the pipe, which must be as short as possible and with at least 5° inclination to permit gas to reach the conservator.

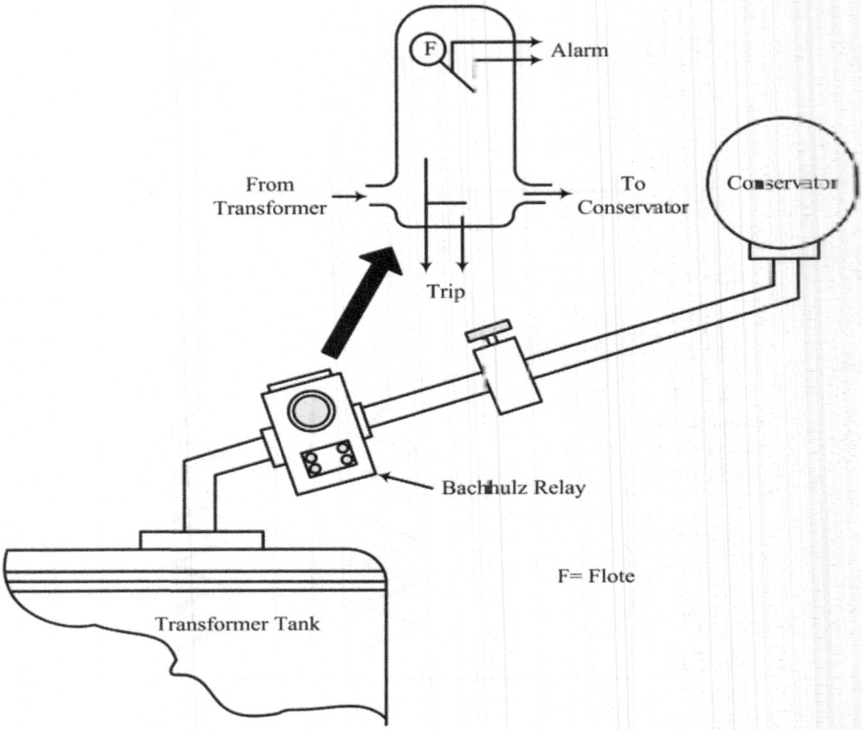

Fig. 1.10 Buchholz relay and its magnified view

1.6.3 *Sudden Pressure Relay*

Sudden pressure relay (SPR) sometimes called a rate of rise of pressure relay. This device detects a rapid rise of pressure than normal pressure [20]. In transformers having a gas cushion instead of a conservator tank, the tripping unit of the Buchholz relay is not applicable and is replaced by a 'sudden pressure relay' which is built into the tank. It has a diaphragm that is deflected by differential oil pressure during the rate of rising of pressure. The gas accumulating unit in such transformers is located at the top of the tank. The relay is set for an operation on a rate of rising in the pressure of 50 g/cm^2/s and a minimum differential gas pressure with 20 g/cm^2/s [7]. Normally, SPR is provided above 5 MVA transformers [21].

1.7 Overall Arrangements of Transformer Protective

For 132 kV and above grid system protective schemes needs high-speed fault clearance for stability point of view and reduce damage due to a fault [3]. For discriminating in a zone and out of zone transformer faults, differential protection and in back up restricted earth fault protection is used. Gas relays fitted with the main transformer and temperature actuator are fitted for alarm and trip to protective schemes. Figure 1.11 illustrates overall protection for a grid-connected transformer with CT/PT wiring and control circuit wiring diagram.

1.8 Past Developments in Transformer Protective Schemes

During the last few decades, astounding success has been achieved in the field of the digital and numerical approach based relaying schemes for the protection of the transformer. Many relaying schemes have been developed by scientists and researchers using microcontrollers, DSPs, AI techniques with hardware narration. These are explained in the following sub-sections.

1.8.1 Adaptive Digital Differential Protection for Transformer

Adaptive digital differential protection is a modified version of the differential protective schemes based on a digital and numerical differential relay. The feature of a biased percentage differential relay is adaptively adjusted as per the requirement, types of the transformer, and severity of protective scheme. So many schemes are elaborated by researchers on many aspects based on CT saturation, inrush, DGA based consideration.

References [22–39] address adaptive digital differential protection. Zhang et al. [22] elaborated self-adaptive transformer differential protection based on a practical solution that is applicable to practice. A major advantage of this scheme is the self-control of characteristics parameters. Adaptive two-stage characteristics on delta-hexagonal type Phase Shifting Transformer (PST) [23] with single-core Differential Current Measuring Principle (DCMP) employed successfully. Performance of the transformer Restricted Earth Fault (REF) relay [24], also improved with a logic of adaptive restraint current technique in a transformer. Kojovic et al. [25] demonstrated innovative differential protection of arc furnace transformers on Rogowski coil sensors. Smith et al. [26] reviewed the concept of coordinating time overcurrent relays with Current Transformer (CT) saturation effects. Alencar et al. [27] have identified inrush currents based on the differential current gradient and have compared schemes with ANN and WT-mathematical morphology. Khan and Sidhu

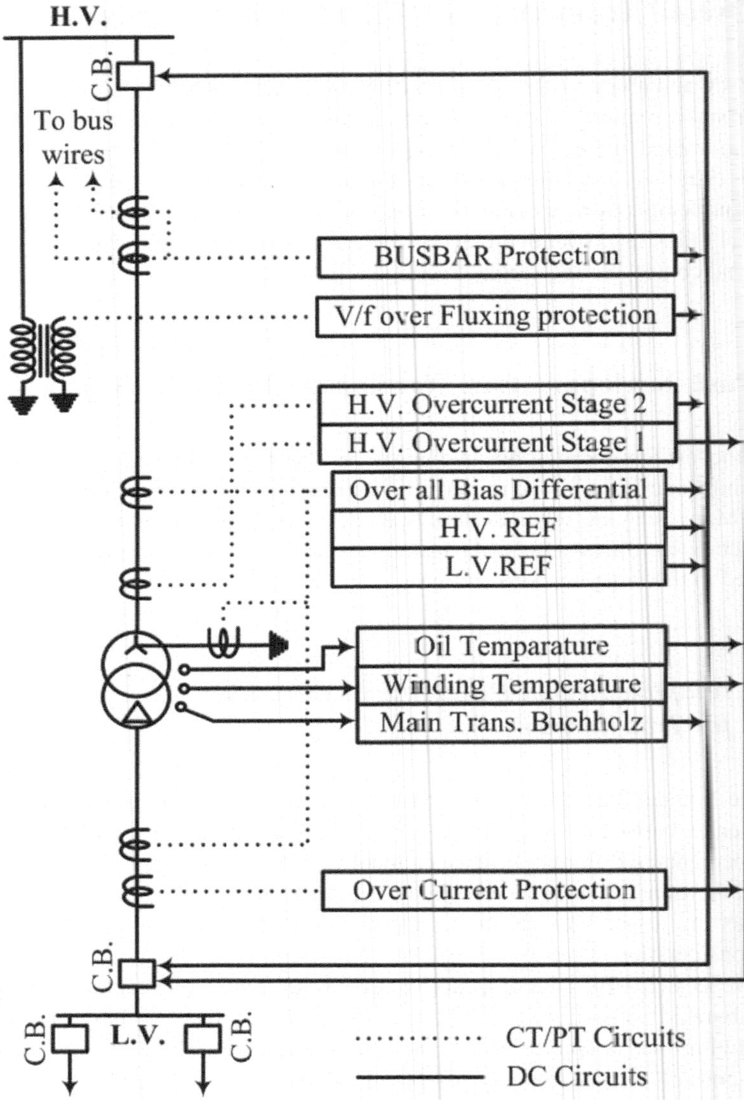

Fig. 1.11 Overall arrangements of protective schemes for typical grid transformer

[28] presented the stabilities of the algorithm under various conditions and minimize the error introduced by the PST series-winding saturation and CT saturation but not nullify. Weight factor-based power transformer protection implemented successfully as a multi-region adaptive differential relay [29]. Based on V–I differential relationships [30] adaptive schemes developed for the protection of standard-delta

phase-shifting transformer. Ingram et al. [31] recognized system-level tests with two slope characteristics. Dmitrenko et al. [32] presented clear digital differential protection of transformer with two-stage digital differential protection. Hajipour et al. [33] proposed CT Saturation and compensation scheme with two-stage (1) Deformed Signal Compensation (DSC) (2) DC offset current compensation (DOCC) for transformer differential protection. Removal of residual flux in transformers [34] proposed by the use of an alternating polarity DC voltage source. Superimposed component comparison [35] based on the internal fault fast identification criterion excellently elaborated. Classification of the internal fault and magnetizing inrush based on Auto-Correlation Function (ACF) [36] explained successfully, speed of response in the proposed algorithm is a half-cycle however ACF itself has a complicated nature. Consideration of system complexity due to CT saturation with adaptive unit type protection applied on distribution transformer with a suitable manner by the third derivative of current based protective schemes [37]. Adaptive real-time monitoring [38, 39] theme based techniques also introduced in transformer protection.

Even though in adaptive digital differential protection average operating time is large, deficiency inefficiency and complicated execution are a major problem. Most of the researchers have not involved nonlinear load conditions, Hence in case of an external fault, schemes may mal-operate. Also, the execution and operating time in the microprocessor-based relaying system depends on the processor speed and capacity of RAM.

1.8.2 DFT, FFT and Other Filtration Based Transformer Protective Schemes

Multi-dimensional DFT and basic FFT are utilized to decompose the signal for further analysis to discriminate fault or normal conditions [40]. Equation 1.4 represents the FFT analysis

$$h[n] * f[n] = \sum_{i=1}^{N} h[n] f[N-n] \tag{1.4}$$

Figure 1.12 shows the block diagram of a Fourier Transform (FT) for transformer protection. CT and PT signals are processed through data acquisition and this data is analyzed with Analogue to Digital Converter (ADC). Then this signal is decomposed with Fourier transform and then compared with a pre-decided threshold value.

Fani et al. [41] proposed method based measurement of second harmonic, wave shape distortion detection for internal fault with impedance and inrush current with 200 experimental cases. DC Decaying components are a major issue in protective schemes, DC component, and harmonic based analysis for detection of inrush current [42] introduced based on filtration technique. Based on the harmonic content of the differential current, Extended Park's vector approach based [43] protective schemes

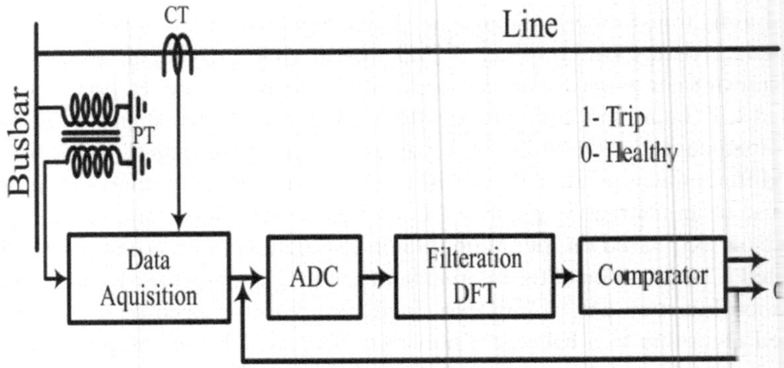

Fig. 1.12 DFT/FFT based algorithm

are introduced with an experimental investigation. Moravej and Abdoos [44] introduce Hyperbolic S-transform (HST) with a discrete version of the S-transform based fault detection scheme. Wagh et al. [45] presented a digital filtering technique based on harmonic analysis and DC component extraction in power transformer protection. Hamilton [46] explained factors affecting on inrush current and harmonic restraint technology. Ma et al. [47] presented a harmonic restrain method using a feature of fundamental current amplitude and compared with 2nd harmonic restrain based algorithm. Again, Ma et al. [48] introduced CT saturation with a combination of generalized morphological filter and grid fractal theory for transformer protection. Moravej et al. [49] identified time-frequency analysis based schemes using hyperbolic S-transform (HST) filter. Khan et al. [50] offered a directional comparison technique based approach for phase-shifting transformer (PST) protection with the DFT based algorithm. However, the additional cost of a voltage transformer and complexity is the main issue. Ashrafian et al. [51] illustrated Time-Time Transform (TT), Short Time Fourier Transform (STFT) for diagnosis of fault in power transformer. To obtain quick and receptive dynamic changes in a differential current gesture using various matrices second harmonic component correlated with intrinsic mode function energy entropy-based technique introduced for traction transformer [52]. Faster, higher accurate with little computational burden, two moving windows based [53] schemes are introduced for identification of an internal fault. Farzad et al. [54] presented secondary fault detection with the help of primary side data using harmonic reduction analysis. Hosny and Sood [55] offered Phasor Amplitude Difference (PAD) based schemes to discriminate of inrush and fault. Stanbury et al. [56] proposed the effect of CT saturation on power transformer protection and various methodologies like 2nd harmonic restrain and Wavelet Transform. Lin et al. [57] explained abnormality detection based on 2nd harmonic restraint and countermeasures on the differential protection of a converter transformer. Bernabeu [58] presented the effects of the harmonic on geometrically induce a current (GIC). Babak et al. [59] proposed an algorithm on extracting the operating segment of the artificial characteristic on a half-cycle data window for online core modeling.

Noshad et al. [60] presented Discrete Wavelet Transform and Clarke's Transform based algorithm for ultra-saturation phenomenon. Saturation index-based CT saturation detection DFT algorithm [61] discriminate high as well as low CT saturation conditions. To overcome the adulteration of DFT based techniques, Kalman filtering based directional transformer protection techniques [62] are also suggested.

The main drawback of the filtering techniques is considerable noise penetration and most of the cases involve the effect of 2nd harmonics. In transformer protection, 2nd harmonic content is not unique because they are generated also under CT saturation and other external fault condition [63]. In some literature, a voltage transformer is also required which increases the cost of protective schemes. Test validation is taken with insufficient data collection even in such cases fault inception angle, high resistance internal fault, type test conditions are not involved. Moreover, the conversation speed of FFT, DFT, and other filtering based techniques for fault detection and classification is not comparable to other techniques.

1.8.3 Sequence Component-Based Transformer Protection Schemes

Effect of positive, negative, and zero sequence components during various types of faults are used as a decision making parameter for protective schemes [64]. The sequence component of the current and voltage based algorithm is explained with a block diagram in Fig. 1.13. Data of voltage and current is collected through data acquisition techniques and then it is converted into sequence component form. This data is compared with a preset threshold value which is considered based on practical as well as theoretical points of view for fault analysis.

The ratio of negative sequence component of primary and secondary side current [65] based technique proposed to the detection of internal fault using FFT. However, this scheme involved only an internal fault. Jenner et al. [66] elaborated gradient vector angle based analysis on the differential current to identify internal fault and inrush condition. Hosny and Sood [55] examined phase angle difference (PAD) based

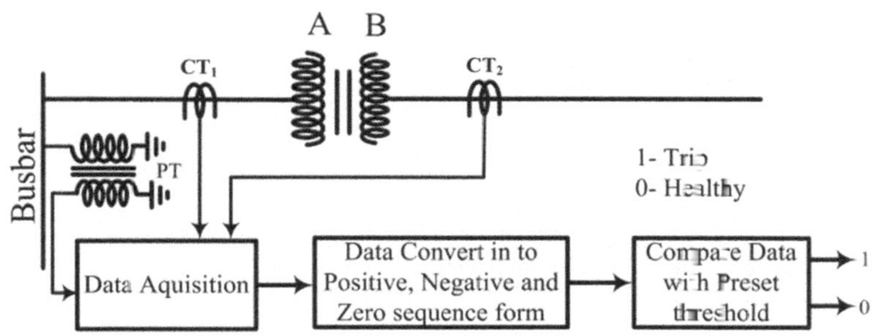

Fig. 1.13 Sequence component-based algorithm

discrimination but the scheme has not considered amplitude comparison. Zacharias and Gokaraju [67] presented a negative sequence component-based turn to turn fault detection techniques using phase and magnitude information. The angle of positive sequence component phasor and magnitude of negative sequence component-based [68] schemes involved for transformer protection. To improve conventional protective schemes of percentage bias differential relay of power transformer included angle of the sequential component-based scheme are incorporated as a parallel path [69]. Same as to improve differential protection phasor angle between primary and secondary currents [70] are also successfully incorporated. The second derivative of differential current is utilized to discriminate inrush conditions in this scheme. However, the operating time of turn fault detection is a major issue.

1.8.4 Artificial Intelligence (AI) Based Transformer Protection Schemes

Artificial Neural Networks (ANN) are logically used in power system protection as they are related to the structure of the human brain. The neurons accomplish the unique feature of ANN structure which can be used to estimate any continuous function. Techniques of Artificial Intelligence (AI) also involve a genetic algorithm and fuzzy system for the protection of the transformer. An equation of neuron for ANN is as follow

$$Y_k = \varphi_k \sum_{i=1}^{m} W_i * X_i \tag{1.5}$$

where $\chi_i =$ input data, $W_i =$ synaptic weights and $\varphi_k =$ activation function.

Block diagram of the Artificial Neural Network (ANN) based transformer differential protection is shown in Fig. 1.14. Differential current samples taken from the secondary of CT are given to ADC through Signal Conditioning (SC) unit and Anti Aliasing Filter (AAF) block. Then this signal is trained and tested by Neural Network

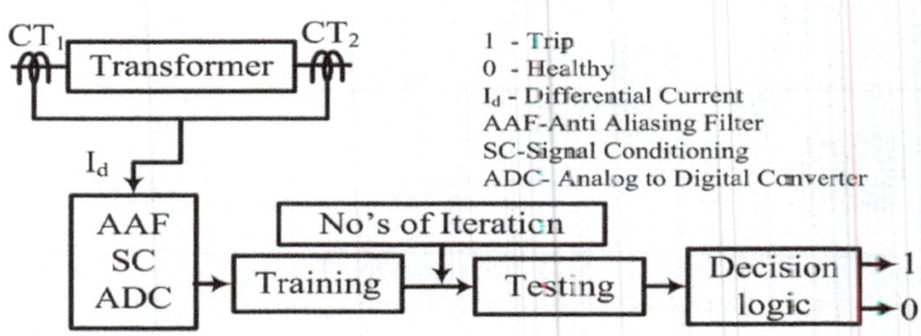

Fig. 1.14 ANN-based algorithm

(NN), and a final decision is taken either for internal fault, external fault, or abnormal conditions.

Mohammad et al. [71] presented ANN with Bayesian Classifier (BC) with swarm base optimization. Sumathi and Bansilal [72] presented detailed ANN for proper coordination of STATCOM between tap changing transformer and generator excitation with a least-square optimization technique. Tripathy et al. [73–75] presented wave shape recognition using neural network principal component analysis (NNPCA) techniques, Optimal Radial Basis Function Neural Network (ORBFNN) and Optimal Probabilistic Neural Network (OPNN) with PSO algorithm. The proposed wave-shape identification based technique is independent of the harmonics content. Moravej et al. [76] estimated a Radial Basis Function (RBF) learning algorithm using two different ANN structures. They again presented ANN [77] as a pattern classifier for power system diagnosis with considerable learning error. Zhalefar and Sanaye-Pasand [78] illustrated an extended blocking scheme based on Fuzzy-logic with the ratio of fundamental to a second harmonic current component. Barbosa et al. [79] proposed a fuzzy system and Clarke's transform with the Mamdani method for mathematical operation. Barbosa et al. [80] elaborated estimation of the current harmonic components by GA, using Shannon's entropy (ANN) CT saturation correction is done and Fuzzy based decision making hybrid systems. Chaiyan et al. [81] explained the PNN based algorithm with a measurement of inductance and resistance to classify an internal fault in a transformer. Ozgonenel et al. [82] implemented ANN as a transformer fault classifier with Wavelet Transform. Vishwakarma et al. [83] introduced a Genetic Algorithm (GA) using trained Master-slave Neural Network (MSNN) where ANN is used for the pattern classifier. Arshad et al. [84] proposed a fuzzy logic technique for condition monitoring and cost-effective optimization techniques for transformer management and decision making. Bejmert et al. [85] offered fuzzy reasoning techniques to limit computational complexity; the simplest membership functions have been employed. Samaher et al. [86] proposed a hybrid methodology of GA with Genetic Neural Computing (GNC) on DGA for prediction and detection of a fault in the transformer. Balaga et al. [87] introduced parallel hidden layered ANN architectures with trained GA for fault discrimination in the transformer.

Till date, many schemes are introduced after research based on ANN and fuzzy, but large training sets, tedious training process, and a large number of neurons are the several disadvantages of the neural network-based schemes. Moreover, the speed of operation, complexity, dependability, accuracy, and security are several limitations of AI techniques. In ANN learning error is considerable even during external fault and under CT saturation conditions possibilities of malfunctioning are higher. Also, the execution of training/testing may not converge as it starts at random and can stop at a local minimum.

1.8.5 Wavelet Transforms (WT) Based Transformer Protection Techniques

A wavelet principle is the same as Fourier analysis, which is useful in image compression and digital signal processing with a mathematical function. It is used to decompose the discrete signal, level by level, with sub-band of frequency to find the rapid change in signal. The effectiveness of the method depends on the mother wavelet selected for the fault analysis. Equation 1.6 represents wavelet filtering by using the wavelet High Pass Filter (HPF) coefficient to a discrete signal.

$$h[n] * f[n] = \sum_{i=1}^{N} h[n] f[N - n] \tag{1.6}$$

where $h_{[n]}$ = HPF coefficient, $f_{[n]}$ = discrete input signal, N = circular window length.

Wavelet Transform based algorithm is illustrated in Fig. 1.15.

Differential Current (I_d) is processed through AAF, SC, and ADC then this signal is decomposed through Wavelet Packet Transform (WPT). Finally, a signal is compared with relay logic and on the basis of result particular fault is classified in a transformer.

Aktaibi et al. [88] offered WPT based hybrid technique on a three-phase stationary coordinate system (abc) to the dq (two-phase) rotating coordinate. Shah [89] proposed Support Vector Machine (SVM) for transformer fault classification through WT as a feature extraction tool. Chaiyan et al. [90] elaborated mother wavelet-based spectrum comparison technique using Discrete Wavelet Transform (DWT). Rahmati and Sanaye-Pasand [91] illustrated pattern recognition based fast WT algorithm for distinguishing inrush and internal fault. Mohammad Hossein et al. [92] presented a discrete wavelet transform based on two indices and by level threshold compression. Ramesh and Sushama [93] depicted a Wavelet Packet Transform (WPT) with energy levels approximation. Medeiros et al. [94] explained Maximal Overlap Discrete Wavelet Transform (MODWT) for differential protection of the transformer. Noshad

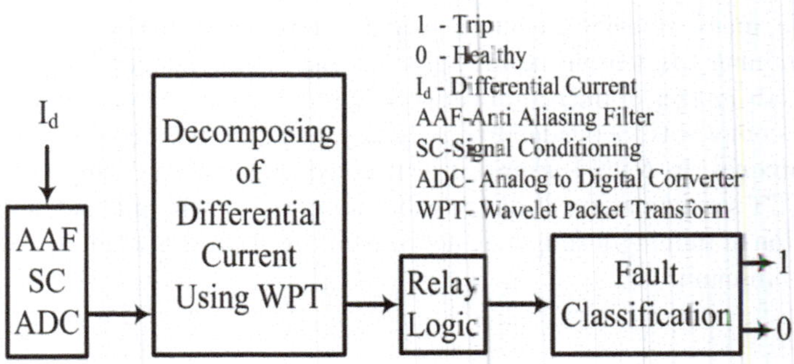

Fig. 1.15 WT based algorithm

et al. [95] presented Clarke's Transform and DWT based technique with Daubechies-4 as a mother wavelet. They have also considered the effect of considering the ultra- saturation phenomenon. Oliveira et al. [96] discovered adaptive differential protection based on transient signal analysis with DWT. Atthapol and Chaiyan [97] suggested DWT with low-frequency components differential current to discriminate inrush, internal fault, and external fault. Maya et al. [98] describes Empirical Wavelet Transform (EWT) and SVM for fault and inrush identification in a transformer.

The Wavelet Transform based schemes require more concentration to select various parameters such as wavelet type, level of decomposition, threshold, and other related parameters. Moreover, the fault identification scheme based on Wavelet requires a high sampling rate of the order of 20–50 kHz. The schemes described above based on WT are good, but some of them are not tested with varying parameters of a transformer and various fault conditions like a fault at the percentage of winding, types of fault, fault inception time, high resistance fault, and variation of load.

1.8.6 Classifier Technique Based Transformer Protection Schemes

Like Support Vector Machine [99], Relevance Vector Machine [100], H-Extreme Learning Machines [101] schemes belong to classifier based protective schemes. Mostly those schemes are applicable where higher accuracy is demanded like the medical field. In classifier techniques algorithms depends on logistic regression, decision tree, vectors. Dataset Source and Contents depend on classes, attributes, instants, and extracted databases. Exploratory Data Analysis depends on various data variables.

For example, the SVM-based block diagram of a protection scheme is shown in Fig. 1.16. Differential current (I_d) is analyzed through AAF, SC, and ADC. This data are first separated in training and testing sets. Initially, the SVM trained and classifier model generated is used for testing of the remaining data set. At last, the classifier will decide the internal or external fault to the transformer.

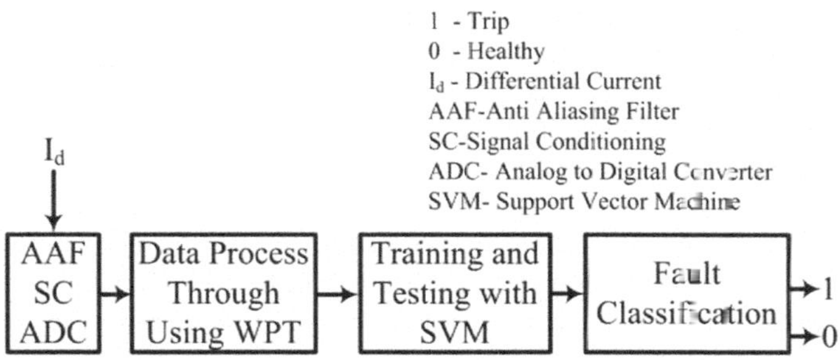

Fig. 1.16 SVM based algorithm

SVM is a classifier method that produces an accurate result with less extrapolation and robust against noise compare to WT based scheme. Wu et al. [102] described the use of LIBSVM for the protection of the transformer with a selection of kernel function which introduces training and testing of fault data. Shah and Bhalja [89] elaborated SVM for classification of fault and disturbances in power transformer compare to Wavelet Transform Spectral Energy (WSE) and Probabilistic Neural Network (PNN). They show a reasonable efficiency of the proposed technique during different inrush and fault conditions. Wei et al. [103] presented Practical Swam Optimization (PSO) based hybrid Least-square SVM (LSSVM) for identification of an incipient fault in a transformer.

SVM puts hyper-plane between two different data classes providing a maximum margin parameter. There is a specific cost function for this kind of model which regulates the plane until the data being successfully classified with minimum error. SVM offers an advantage over ANN that it has a simple geometric interpretation and gives a sparse solution.

However, apart from a binary classifier method, SVM has a lack of transparency in outcome for multiclass, pair-wise classification methodology. Due to the main requirement of satisfaction of Mercer's condition for SVM kernel predictions are not probabilistic. With the use of general-purpose kernels with model search and cross-validation provides insufficient results as they don't take peculiarities of the training data into account. Whereas, satisfying Mercer's theorem in SVM means classifier must have a positive semi-definite convex function. This guarantees the existence of an underlying map allowed us to select kernels such that the underlying map could be Gaussian, Sigmoid Kernels.

1.8.7 All Other Methodology Used for Transformer Protection

Other than fundamental technology used so far, various techniques are involved with specific limitations in the field of transformer protection. Wave shape properties [104, 105] based schemes are explained by Hooshyar and Sanaye-Pasand. Matrix-based algorithm [106–108] presented with model base analysis and compared it with various techniques. In these schemes, types of testing data are very less for validation of the technique. Kang et al. [109] presented an Incremental Flux Linkages based method for fault detection in the transformer. Lei et al. [110] expressed an improved lumped circuit based on the transfer function. But the loss in the insulation and conductors are not considered. Tian et al. [52] presented chaos theory with energy entropy and correlation dimension based intrinsic mode function (IMF). For detection of ultra saturation in transformer fourth-order Runge–Kutta scheme [111] incorporated. However, there is a possibility of false operation due to residual flux and fault inception angle. Behjat et al. [112] identify external circuit equations with transient finite element method coupling to analyze the transient behavior with

Maxwell's equations. Schettino et al. [113] presented the Sound to Noise Ratio (SNR) method for saturation detection. Sine-wave least-squares curve fitting method (SCF) [114] used to prejudiced internal fault from magnetizing inrush current. Lissajous graphical analysis [115] of voltage and current based on winding deformations for online monitoring. Rudez and Mihalic [116] presented Eigen-values and Eigen-vectors based Sympathetic inrush current effect. However, the main drawback is that they have not measured residual flux which may operate the scheme. Dashti et al. [117] suggested a morphological gradient (MG) based mathematical morphology scheme for discriminating large inrush currents from fault current. But, this scheme is validated with less number of testing data set. Now a day combination of real-time monitoring and adaptive protection [38] of the power transformer is also given its prime importance due to the highly burning importance of transformer in the power system.

Even though, numbers of transformer protection schemes have been suggested up to now there survives a lot of prospects for advanced enlargement particularly on the efficient discrimination among in-zone and out of zone transformer fault and fault classification with other external abnormalities. Hence, in the forthcoming section, a concept of a new digital transformer protection scheme has been offered.

1.9 Combined Filtration and Classification Scheme for Transformer Protection

Recently many protective schemes are utilized to discriminate internal and external fault on bases of microcontroller, DSP, ANN, WT, SVM, travelling waves, and mathematical morphology, with software simulation and hardware implementation. They are explained by researchers in literature as described in the previous section. However, each and every methodology has its own benefit and limitations while implementing in real-time applications.

Recently, the combination of filtration technique and classification tool has been used in the power system for fault classification [118]. It is true that whenever a fault occurs, the knowledge of only fundamental component information may not be sufficient for secured fault discrimination. It is thus necessary to pre-process the input data and extract useful features for training using DFT, Kalman filter and WPT. In combination of these feature bagging, the classification tool used are ANN, PNN, SVM, ELM and RVM to discriminate internal fault, external fault, and various abnormal conditions like CT saturation, magnetizing inrush, and high resistance internal fault with varying power system parameters. SVM and RVM are better option with respect to time consumption and fault classification accuracy [119].

1.10 Summary

A transformer is the heart of the power system, so mal-functioning in a protective scheme generates numerous problems and gives the worst effect on power system stability. This article presents a literature review with past methodology to detect and classify various faults in power transformers. Survey of the various methodology and concepts of transformer protection is carried out with proper relevant background, the actual requirement of field, past events, and current scenarios with consideration of future requirements. This piece of writing is based on the work presented in many research articles published in the last 30 years and periodic bibliographic. After reviewing all methodology for transformer protection, authors have concluded that techniques based on ANN, GA, Fuzzy, AI, WT, and SVM are the most efficient techniques but having some constraint and limitation. An efficient and reliable relaying scheme for transformer protection will be derived with the combination of techniques mentioned this chapter. Further extension of work in this procession is carried out in Chap. 5 [100] and Chap. 6 [101].

References

1. Rajurkar SS, Nandapurkar JG, Kulkarni A (2010) Analysis of power transformer failure in transmission utilities. In: 16th national power system conference, pp 180–185
2. Binder W. Trends in power transformer failure analysis. Transformers Mag 1(1) 30–33
3. Rushton J (1995) Power system protection, 2nd edn. The Institutes of Electrical Engineers, London, United Kingdom (Revised by Mewes KG)
4. Paithankar YG, Bhide SR (2010) Fundamentals of power system protection. PHI Learning Pvt. Ltd., New Delhi, India
5. Rao S (2008) Switchgear and protection, 1st edn. PHI Learning Pvt. Ltd., New Delhi, India
6. IEEE Guide for Protecting Power Transformers (Revision of IEEE Std C37.91-2000) (2008) IEEE Power Engineering Society Sponsored by the Power System Relaying Committee New York, USA
7. Anderson PM (1999) Power system protection, 1st edn. A John Wiley & Sons Inc. Publication, IEEE Press, New York, pp 673–709
8. Van AR, Warrington C (1962) Protective relays—their theory and practice, 1st edn, vol 1. Chapman and Hall, London
9. IIAS Industrial and CPS Committee (1986) IEEE recommended practice for protection and coordination of industrial and commercial power systems: approved 19 Sept 1985, Reaffirmed 27 June 1991 IEEE Standards Board: approved 28 Feb 1986, Reaffirmed 9 Dec 1991. American National St. IEEE
10. Elmore WA (2003) Protective relaying theory and applications, 2nd edn. Marcel Dekker, New York
11. Wu QH, Lu Z, Ji TY (2009) Protective relaying of power systems using mathematical morphology, 1st edn. Springer, London
12. Soman SA. Power system protection learning resources. NPTEL, Web Course, IIT, Bombay, India
13. Blackburn JL, Domin TJ (2015) Protective relaying: principles and applications, 3rd edn. CRC Press, Tailor and Francies Group, New York
14. Brownlee WR (1944) Transformer magnetizing inrush currents and influence on system operation. Trans Am Inst Electr Eng 63(6):423–500

15. Blume LF, Camilli G, Farnham SB, Peterson HA (1944) Transformer magnetizing inrush currents and influence on system operation. Trans Am Inst Electr Eng 63(6):366–375
16. Holcomb JE (1961) Distribution transformer magnetizing inrush current. Trans Am Inst Electr Eng Part III Power Apparatus Syst 80(3):697–702
17. Kulkarni SV, Khaparde SA (2004) Transformer engineering: design and practice. Taylor & Francis
18. Patel DD, Mistry KD, Chothani NG (2017) Transformer inrush/internal fault identification based on average angle of second order derivative of current. In Asia-Pacific power and energy engineering conference, APPEEC, pp 1–6
19. Raichura MB, Chothani NG, Patel DD, Mistry KD (2017) Analysis of various factor affecting on saturation of current transformer used in power system protection. ADIT J Eng 14(1):29–34
20. Ram B (2011) Power system protection and switchgear. Tata McGraw-Hill Education, New Delhi, India
21. Bhalja B, Maheshwari RP, Chothani NG (2017) Protection and switchgear, 2nd edn. Oxford University Press, New Delhi, India
22. Zhang W, Tan Q, Miao S, Zhou L, Liu P (2013) Self-adaptive transformer differential protection. IET Gener Transm Distrib 7(1):61–68
23. Khan U, Sidhu TS (2013) A comparative performance analysis of three differential current measuring principles associated with phase shifting transformer protection. In: 2013 12th international conference on environment and electrical engineering, pp 297–302
24. Davarpanah M, Sanaye-Pasand M, Iravani R (2013) Performance enhancement of the transformer restricted earth fault relay. IEEE Trans Power Deliv 28(1):467–474
25. Kojovic LA, Bishop MT, Sharma D (2013) Innovative differential protection of power transformers using low-energy current sensors. IEEE Trans Ind Appl 49(5):1971–1978
26. Smith T, Hunt R (2013) Current transformer saturation effects on coordinating time interval. IEEE Trans Ind Appl 49(2):825–831
27. Alencar RJN, Bezerra UH, Ferreira AMD (2014) A method to identify inrush currents in power transformers protection based on the differential current gradient. Electr Power Syst Res 111:78–84
28. Sidhu TS (2014) New algorithm for the protection of delta-hexagonal phase shifting transformer. IET Gener Transm Distrib 8(1):178–186
29. Dashti H, Sanaye-Pasand M (2014) Power transformer protection using a multiregion adaptive differential relay. IEEE Trans Power Deliv 29(2):777–785
30. Khan UN, Sidhu TS (2014) Protection of standard-delta phase shifting transformer using terminal currents and voltages. Electr Power Syst Res 110:31–38
31. Ingram DME, Schaub P, Taylor RR, Campbell DA (2014) System-level tests of transformer differential protection using an IEC 61850 process bus. IEEE Trans Power Deliv 29(3):1382–1389
32. Dmitrenko AM, Zhuravlev DP (2015) An analysis and investigation of methods of braking digital differential protections of block transformers. Power Technol Eng 49(1):72–77
33. Hajipour E, Vakilian M, Sanaye-Pasand M (2015) Current-transformer saturation compensation for transformer differential relays. IEEE Trans Power Deliv 30(5):2293–2302
34. de Leon F, Farazmand A, Jazebi S, Deswal D, Levi R (2015) Elimination of residual flux in transformers by the application of an alternating polarity DC voltage source. IEEE Trans Power Deliv 30(4):1727–1734
35. Lin X, Lu J, Zhang R, Tong N, Adio OS, Li Z (2015) Internal fault fast identification criterion based on superimposed component comparison for power transformer. Int J Electr Power Energy Syst 73:491–497
36. Samet H, Ghanbari T, Ahmadi M (2015) An auto-correlation function based technique for discrimination of internal fault and magnetizing inrush current in power transformers. Electr Power Compon Syst 43(4):399–411
37. Patel DD, Chothani N, Mistry KD, Tailor D (2018) Adaptive algorithm for distribution transformer protection to improve smart grid stability. Int J Emerg Electr Power Syst 19(5):1–14

38. Chothani NG, Raichura MB, Patel DD, Mistry KD (2019) Real-time monitoring protection of power transformer to enhance smart grid reliability. Electr Control Commun Eng 15(2):104–112
39. Chothani NG, Raichura MB, Patel DD, Mistry KD (2018) Real-time monitoring protection of power transformer to enhance smart grid reliability. In: 2018 IEEE electrical power and energy conference (EPEC), pp 1–6
40. Bhide SR (2014) Digital power system protection, 1st edn. PHI Learning Private Limited, Delhi
41. Fani B, Golshan MEH, Abyaneh HA, Saghaian-nejad M (2011) A runs test-based method for discrimination between internal faults and inrush currents in power transformers. Eur Trans Electr Power 21(3):1392–1408
42. Madani SM, Rostami M, Gharehpetian GB, Haghmaram R (2012) Inrush current limiter based on three-phase diode bridge for Y-yg transformers. IET Electr Power Appl 6(6):345–352
43. Oliveira LMR, Cardoso AJM (2012) Extended Park's vector approach-based differential protection of three-phase power transformers. IET Electr Power Appl 6(8):463–472
44. Moravej Z, Abdoos AA (2012) An improved fault detection scheme for power transformer protection. Electr Power Compon Syst 40(10):1183–1207
45. Wagh SR, Kumar S, Sreeram V (2013) Extraction of DC component and harmonic analysis for protection of power transformer. In: 2013 IEEE 8th conference on industrial electronics and applications (ICIEA), pp 32–37
46. Hamilton R (2013) Analysis of transformer inrush current and comparison of harmonic restraint methods in transformer protection. IEEE Trans Ind Appl 49(4):1890–1899
47. Ma J, Wang Z, Zheng S, Wang T, Yang Q (2013) A new algorithm to discriminate internal fault current and inrush current utilizing feature of fundamental current. Can J Electr Comput Eng 36(1):26–31
48. Ma J, Gao X, Wang Z, Yang Q (2013) A new adaptive method to identify TA saturation using grille fractal. Can J Electr Comput Eng 36(4):152–157
49. Moravej Z, Abdoos AA, Sanaye-Pasand M (2013) Power transformer protection scheme based on time-frequency analysis. Int Trans Electr Energy Syst 23(4):473–493
50. Khan UN, Sidhu TS (2014) A phase-shifting transformer protection technique based on directional comparison approach. IEEE Trans Power Deliv 29(5):2315–2323
51. Ashrafian A, Vahidi B, Mirsalim M (2014) Time-time-transform application to fault diagnosis of power transformers. IET Gener Transm Distrib 8(6):1156–1167
52. Li Y, Tian X, Li X (2014) Hybrid algorithm for traction transformer differential protection based on intrinsic mode function energy entropy and correlation dimension. IET Gener Transm Distrib 8(7):1274–1283
53. Abniki H, Sanaye-Pasand M (2014) A novel technique for internal fault detection of power transformers based on moving windows. Int Trans Electr Energy Syst 24(9):1263–1278
54. Dehghan Marvasti F, Samet H (2014) Fault detection in the secondary side of electric arc furnace transformer using the primary side data. Int Trans Electr Energy Syst 24(10):1419–1433
55. Hosny A, Sood VK (2014) Transformer differential protection with phase angle difference based inrush restraint. Electr Power Syst Res 115:57–64
56. Stanbury M, Djekic Z (2015) The impact of current-transformer saturation on transformer differential protection. IEEE Trans Power Deliv 30(3):1278–1287
57. Lu J, Lin X et al (2015) Abnormal operation behavior analysis and countermeasures on the differential protection of converter transformer. Int J Electr Power Energy Syst 64:516–525
58. Bernabeu EE (2015) Single-phase transformer harmonics produced during geomagnetic disturbances: theory, modeling, and monitoring. IEEE Trans Power Deliv 30(3):1323–1330
59. Khalkhali B, Sadeh J (2015) Transformer differential protection by online core modeling and orthogonal polynomials. IEEE Trans Power Deliv 30(5):2146–2153
60. Noshad B, Ghanavati B, Ahmadzadeh M, Mohammadzadeh S (2015) Control of unusual mal-operation of three-phase power transformer differential protection due to ultra saturation phenomenon based on Clarke's Transform and Discrete Wavelet Transform. Renew Sustain Energy Rev 51:1276–1287

61. Patel DD, Mistry KD, Chothani NG (2016) Digital differential protection of power transformer using DFT algorithm with CT saturation consideration. In: 2016 national power systems conference (NPSC), pp 1–6
62. Patel DD, Mistry KD, Raichura MB, Chothani N (2018) Three State Kalman filter based directional protection of power transformer. In: 20th national power systems conference (NPSC), pp 1–6
63. Pihler J, Grcar B, Dolinar D (1997) Improved operation of power transformer protection using artificial neural network. IEEE Trans Power Deliv 12(3):1128–1136
64. Chothani N, Patel D, Raichura M (2019) Transformer protection with sequence components and digital filters, 1st edn. LAP LAMBERT Academic Publishing, Latvia, European Union
65. Vahedi A, Behjat V (2011) Online monitoring of power transformers for detection of internal winding short circuit faults using negative sequence analysis. Eur Trans Electr Power 21(1):196–211
66. de Alencar RJN, Bezerra UH (2013) Power transformer differential protection through gradient of the differential current. J Control Autom Electr Syst 24(1):162–173
67. Zacharias D, Gokaraju R (2016) Prototype of a negative sequence turn-to-turn fault detection scheme for transformers. In: 2016 IEEE power and energy society general meeting (PESGM), p 1
68. Patel DD, Chothani NG, Mistry KD (2015) Sequence component of currents based differential protection of power transformer. In: 12th IEEE international conference electronics, energy, environment, communication, computer, control: (E3-C3), INDICON 2015, pp 1–6
69. Patel DD, Mistry KD, Chothani NG (2015) A novel approach to transformer differential protection using sequence component based algorithm. J CPRI 11(3):517–528
70. Patel D, Chothani N, Mistry K (2018) Discrimination of inrush, internal, and external fault in power transformer using phasor angle comparison and biased differential principle. Electr Power Compon Syst 46(7):788–801
71. Yazdani-Asrami M, Taghipour-Gorjikolaie M, Razavi SM, Gholamian SA (2015) A novel intelligent protection system for power transformers considering possible electrical faults, inrush current, CT saturation and over-excitation. Int J Electr Power Energy Syst 64:1129–1140
72. Sumathi S, Bansilal (2015) Artificial Neural Network for coordinated control of STATCOM, generator excitation and tap changing transformer. Int J Electr Power Energy Syst 64:536–541
73. Tripathy M (2009) Neural Network Principal Component Analysis based power transformer differential protection. In: 2009 International Conference on Power Systems, pp 1–6
74. Tripathy M, Ala S (2009) Optimal radial basis function neural network power transformer differential protection. In: 2009 IEEE Bucharest PowerTech, pp 1–8
75. Tripathy M, Maheshwari RP, Verma HK (2010) Power transformer differential protection based on optimal probabilistic neural network. IEEE Trans Power Deliv 25(1):102–112
76. Moravej Z, Vishwakarma DN, Singh SP (2002) Protection and conditions monitoring of power transformer using ANN. Electr Power Compon Syst 30(3):217–231
77. Moravej Z, Vishwakarma DN (2000) Singh PS (2000) ANN-based protection scheme for power transformer. Electr Mach Power Syst 28(9):875–884
78. Zhalefar F, Sanaye-Pasand M (2010) A new fuzzy-logic-based extended blocking scheme for differential protection of power transformers. Electr Power Compon Syst 38(6):675–694
79. Barbosa D, Netto UC, Coury DV, Oleskovicz M (2011) Power transformer differential protection based on Clarke's Transform and fuzzy systems. IEEE Trans Power Deliv 26(2):1212–1220
80. Barbosa D, Coury DV, Oleskovicz M (2012) New approach for power transformer protection based on intelligent hybrid systems. IET Gener Transm Distrib 6(10):1009–1018
81. Jettanasen C, Ngaopitakkul A (2013) A novel probabilistic neural network-based algorithm for classifying internal fault in transformer windings. IEEJ Trans Electr Electron Eng 8(2):123–131
82. Ozgonenel O, Onbilgin G, Kocaman C (2005) Transformer protection using the wavelet transforms. Ondokus Mayis Univ Turk J Electr Eng 13:60–67

83. Vishwakarma DN, Balaga H, Nath H (2014) Application of genetic algorithm trained master slave neural network for differential protection of power transformer. In: 2014 9th international conference on computer engineering systems (ICCES), pp 164–169
84. Arshad M, Islam SM, Khaliq A (2014) Fuzzy logic approach in power transformers management and decision making. IEEE Trans Dielectr Electr Insul 21(5):2343–2354
85. Bejmert D, Rebizant W, Schiel L (2014) Transformer differential protection with fuzzy logic based inrush stabilization. Int J Electr Power Energy Syst 63:51–63
86. Al-Janabi S, Rawat S, Patel A, Al-Shourbaji I (2015) Design and evaluation of a hybrid system for detection and prediction of faults in electrical transformers. Int J Electr Power Energy Syst 67:324–335
87. Balaga H, Gupta N, Vishwakarma DN (2015) GA trained parallel hidden layered ANN based differential protection of three phase power transformer. Int J Electr Power Energy Syst 67:286–297
88. Narendra K, Fedirchuk D, Zhang N, Midence R, Perera N, Sood V (2011) Phase angle comparison and differential rate of change methods used for differential protection of busbars and transformers. In: 2011 IEEE electrical power and energy conference, pp 134–139
89. Shah AM, Bhalja BR (2013) Discrimination between internal faults and other disturbances in transformer using the support vector machine-based protection scheme. IEEE Trans Power Deliv 28(3):1508–1515
90. Jettanasen C, Ngaopitakkul A (2012) The spectrum comparison technique of DWT for discriminating between external fault and internal fault in power transformer. J Int Counc Electr Eng 2(3):302–308
91. Rahmati A, Sanaye-Pasand M (2012) A fast WT-based algorithm to distinguish between transformer internal faults and inrush currents. Eur Trans Electr Power 22(4):471–490
92. Zendehdel MH, Sanaye-Pasand M (2012) Development of two indices based on discrete wavelet transform for transformer differential protection. Eur Trans Electr Power 22(8) 1078–1092
93. Ramesh K, Sushama M (2014) Power transformer protection using wavelet packet transform. In: 2014 international conference on computer communication and informatics
94. Medeiros RP, Costa FB, Fernandes JF (2014) Differential protection of power transformers using the wavelet transform. In: 2014 IEEE PES general meeting|conference exposition, pp 1–5
95. Noshad B, Razaz M, Seifossadat SG (2014) A new algorithm based on Clarke's Transform and Discrete Wavelet Transform for the differential protection of three-phase power transformers considering the ultra-saturation phenomenon. Electr Power Syst Res 110:9–24
96. Oliveira MO, Bretas AS, Ferreira GD (2014) Adaptive differential protection of three-phase power transformers based on transient signal analysis. Int J Electr Power Energy Syst 57 366–374
97. Ngaopitakkul A, Jettanasen C (2014) A discrete wavelet transform approach to discriminating among inrush current, external fault, and internal fault in power transformer using low-frequency components differential current only. IEEJ Trans Electr Electron Eng 9(3):302–314
98. Maya P, Shree SV, Roopasree K, Soman KP (2015) Discrimination of internal fault current and inrush current in a power transformer using empirical wavelet transform. Procedia Technol 21:514–519
99. Chothani NG, Patel DD, Mistry KD (2017) Support vector machine based classification of current transformer saturation phenomenon. J Green Eng River Publ 7:25–42
100. Patel D, Chothani NG, Mistry KD, Raichura M (2018) Design and development of fault classification algorithm based on relevance vector machine for power transformer. IET Electr Power Appl 12(4):557–565
101. Raichura MB, Chothani NG, Patel DD (2020) Identification of internal fault against external abnormalities in power transformer using hierarchical ensemble extreme learning machine technique. IET Sci Meas Technol 14(1):111–121

102. Wu H, Fu C, Guo H, Chen CZ (2010) The research of the fault types recognition in transformer by differential protection based on LIBSVM. In: 2010 3rd international conference on biomedical engineering and informatics, vol 7, pp 2999–3003
103. Wei CH, Tang WH, Wu QH (2014) A hybrid least-square support vector machine approach to incipient fault detection for oil-immersed power transformer. Electr Power Compon Syst 42(5):453–463
104. Hooshyar A, Sanaye-Pasand M (2015) Waveshape recognition technique to detect current transformer saturation. IET Gener Transm Distrib 9(12):1430–1438
105. Hooshyar A, Sanaye-Pasand M (2011) CT saturation detection based on waveform analysis using a variable-length window. IEEE Trans Power Deliv 26(3):2040–2050
106. Zheng T, Chen PL, Qi Z, Terzija V (2014) A novel algorithm to avoid the maloperation of UHV voltage-regulating transformers. IEEE Trans Power Deliv 29(5):2146–2153
107. Noshad B, Tabatabaee S, Ghanavati B, Mohammadzadeh S (2015) A new model to study ultra-saturation phenomenon during the energization of a loaded three-phase power transformer and its effects on differential protection. Int Trans Electr Energy Syst 25(12):3681–3703
108. Eissa MM, Shehab-Eldin EH, Masoud ME, Abd-Elatif AS (2012) Laboratory investigation for power transformer protection technique based on positive sequence admittance approach. Eur Trans Electr Power 22(2):253–270
109. Kang YC, Lee BE, Zheng TY, Kim YH, Crossley PA (2010) Protection, faulted phase and winding identification for the three-winding transformer using the increments of flux linkages. IET Gener Transm Distrib 4(9):1060–1068
110. Lei X, Li J, Wang Y, Mi S, Xiang C (2014) Simulative and experimental investigation of transfer function of inter-turn faults in transformer windings. Electr Power Syst Res 107:1–8
111. Noshad B, Razaz M, Seifossadat SG (2013) A model for the ultra-saturation phenomenon during energization of an unloaded power transformer and its effect on differential protection. Electr Power Compon Syst 41(12):1129–1145
112. Behjat V, Vahedi A (2013) Analysis of internal winding short circuit faults in power transformers using transient finite element method coupling with external circuit equations. Int J Numer Model Electron Netw Dev Fields 26(5):425–442
113. Schettino BM, Duque CA, Silveira PM, Ribeiro PF, Cerqueira AS (2014) A new method of current-transformer saturation detection in the presence of noise. IEEE Trans Power Deliv 29(4):1760–1767
114. Ahmadi M, Samet H, Ghanbari T (2015) Discrimination of internal fault from magnetising inrush current in power transformers based on sine-wave least-squares curve fitting method. IET Sci Meas Technol 9(1):73–84
115. Yao C, Zhao Z, Mi Y, Li C, Liao Y, Qian G (2015) Improved online monitoring method for transformer winding deformations based on the Lissajous graphical analysis of voltage and current. IEEE Trans Power Deliv 30(4):1965–1973
116. Rudez U, Mihalic R (2016) Sympathetic inrush current phenomenon with loaded transformers. Electr Power Syst Res 138:3–10
117. Dashti H, Davarpanah M, Sanaye-Pasand M, Lesani H (2016) Discriminating transformer large inrush currents from fault currents. Int J Electr Power Energy Syst 75:74–82
118. Tipping ME (2001) Sparse Bayesian learning and the relevance vector machine. J Mach Learn Res 1:211–244
119. Chothani NG, Bhalja BR (2016) A new algorithm for busbar fault zone identification using relevance vector machine. Electr Power Compon Syst 44(2):193–205

Chapter 2
CT Saturation Detection and Compensation Algorithm

This division presents a new Current Transformer (CT) saturation detection and compensation algorithm. The proposed algorithm depends on a saturation detection index (Dn) which is derived using derivatives of current signals and Newton's backward difference formulas. The calculated index is continuously compared with an adaptive threshold (Th) to estimate the start and endpoint of CT saturation. A low-pass first-order Butterworth filter is used to suppress noise and harmonics which may present in CT secondary current. The proposed saturation detection algorithm has been tested by considering different values of remanent flux, fault type, fault inception angle, burden resistance, decaying DC component of fault current, and noise. At the same time, MDFT based compensating algorithm has also been proposed to reconstruct the saturated samples. Validation of the proposed scheme is also carried out on a developed laboratory prototype. A comparative evaluation of the proposed algorithm is also carried out with existing schemes. Series of test results from simulation software and laboratory prototype show the effectiveness of the proposed scheme.

Though the main function of the CT is to transform the maximum possible current during normal as well as faulty conditions, its saturation is inevitable. The amount of saturation depends on fault current magnitude, remanence flux, DC component, the time constant of CT, and burden on the secondary side of CT [1, 2]. Several methods have been suggested by researchers for the detection of CT saturation. A new Current Transformer (CT) saturation detection and compensation algorithm proposed based on the derivative technique. The proposed algorithm depends on a saturation detection index (Dn) which is derived using derivatives of current signals and Newton's backward difference formulas. The calculated index is continuously compared with an adaptive threshold (Th) to estimate the start and endpoint of CT saturation. A low-pass first-order Butterworth filter is used to suppress noise and harmonics which may present in CT secondary current. The proposed saturation detection algorithm has been tested by considering different values of remanent flux, fault type, fault inception angle, burden resistance, decaying DC component of fault

D. Patel and N. Chothani, *Digital Protective Schemes for Power Transformer*, Power Systems, https://doi.org/10.1007/978-981-15-6763-6_2

current, and noise. At the same time, MDFT based compensating algorithm has also been proposed to reconstruct the saturated samples. Validation of the proposed scheme is also carried out on a developed laboratory prototype. A comparative evaluation of the proposed algorithm is also carried out with existing schemes. Series of test results from simulation software and laboratory prototype show the effectiveness of the proposed scheme. The proposed scheme has been tested by generating various saturation cases on the CT model available in PSCAD/EMTDC software package [3]. Subsequently, the same algorithm has also been validated by developing a test bench of CT in a laboratory environment.

2.1 Proposed Method for CT Saturation Detection

2.1.1 Proposed Algorithm

The primary current i_1(t) during the transient analysis of CT can be given by [4],

$$i_1(t) = \mathrm{I_{max}}\left[\cos(\omega t - \theta) - e^{-t/\mathrm{T}_P}\cos\theta\right] \quad \text{for t} \geq 0 \tag{2.1}$$

where $\mathrm{I_{max}}$ is the peak value of sinusoidal steady-state fault current, $\mathrm{T_P}$ is the primary time-constant and θ is the fault initiation angle.

The secondary current of CT is given by Eq. (2.2).

$$\mathrm{i_2(t) = Ae}^{-t/\mathrm{T}_S} + \mathrm{Be}^{-t/\mathrm{T}_P} - \mathrm{C} * \sin(\omega \mathrm{t} - \theta - \varphi) \tag{2.2}$$

where, $\mathrm{T_P}$ and $\mathrm{T_S}$ are primary and secondary time constant, respectively, and A & B are constants. In Eq. (2.2), the first and second exponential terms decay with the time constants $\mathrm{T_S}$ and $\mathrm{T_P}$, respectively, whereas the magnitude of the sinusoidal term is given by,

$$C = I_{\max}\omega \mathrm{T_S}\cos\varphi = I_{\max}\sin\varphi = I_{\max}\frac{\omega \mathrm{T_S}}{\sqrt{1 + (\omega \mathrm{T_S})^2}} \quad \text{where,} \quad \tan\varphi = \omega T_S \tag{2.3}$$

The discrete-time version of $\mathrm{i_2}$ (t) is obtained by considering t = nH.

$$i_{2[n]} = \mathrm{Ae}^{-nH/\mathrm{T}_S} + \mathrm{Be}^{-nH/\mathrm{T}_P} - \mathrm{C} * \sin\left(\frac{2\pi}{\mathrm{N}}n - \theta - \varphi\right) \tag{2.4}$$

where H is the sampling interval, N is the number of samples per cycle and n is the recent sample.

The first difference of $\mathrm{i}_{2[n]}$ is given by Eq. (2.5).

$$
\begin{aligned}
\nabla_n^1 &= i_{2[n]} - i_{2[n-1]} \\
&= A(1 - e^{(H/T_S)}) * e^{-(nH/T_S)} + B(1 - e^{(H/T_P)}) * e^{-(nH/T_P)} \\
&\quad - C\left(2\sin\frac{\pi}{N}\right)\sin\left(\frac{2\pi}{N}n - \theta - \varphi - \frac{\pi}{N} + \frac{\pi}{2}\right)
\end{aligned} \tag{2.5}
$$

If the sampling rate is 4 kHz (80 samples per cycle) for a power system frequency of 50 Hz, the sampling interval H = 0.25 ms. By considering $T_S = 1$ s and $T_P = 0.02$ s, the value of $\left(1 - e^{(H/T_S)}\right)$ and $\left(1 - e^{(H/T_P)}\right)$ are exponentially reduced to 0.00025 and 0.0125, respectively [5, 6]. This indicates that the exponential terms ∇_n^1 are considerably reduced and have negligible values since the time constants are large. These values are further reduced for CTs of higher protection class as the second time constant of such CTs are in the range of 3–10 s [6].

The following equations can be derived for the second, third & fourth difference of the CT secondary current.

$$
\begin{aligned}
\nabla_n^2 &= \nabla_n^1 - \nabla_{n-1}^1 \\
&= i_{2[n]} - 2i_{2[n-1]} + i_{2[n-2]} \\
&= A(1 - e^{(H/T_S)})^2 e^{-(nH/T_S)} + B(1 - e^{(H/T_P)})^2 e^{-(nH/T_P)} \\
&\quad - C\left(2\sin\frac{\pi}{N}\right)^2 \sin\left(\frac{2\pi}{N}n - \theta - \varphi - \frac{2\pi}{N}\right)
\end{aligned} \tag{2.6}
$$

$$
\begin{aligned}
\nabla_n^3 &= \nabla_n^2 - \nabla_{n-1}^2 \\
&= i_{2[n]} - 3i_{2[n-1]} + 3i_{2[n-2]} - i_{2[n-3]} \\
&= A(1 - e^{(H/T_S)})^3 e^{-(nH/T_S)} + B(1 - e^{(H/T_P)})^3 e^{-(nH/T_P)} \\
&\quad - C\left(2\sin\frac{\pi}{N}\right)^3 \sin\left(\frac{2\pi}{N}n - \theta - \varphi - \frac{3\pi}{N} + \frac{\pi}{2}\right)
\end{aligned} \tag{2.7}
$$

$$
\begin{aligned}
\nabla_n^4 &= \nabla_n^3 - \nabla_{n-1}^3 \\
&= i_{2[n]} - 4i_{2[n-1]} + 6i_{2[n-2]} - 4i_{2[n-3]} + i_{2[n-4]} \\
&= A(1 - e^{(H/T_S)})^4 e^{-(nH/T_S)} + B(1 - e^{(H/T_P)})^4 e^{-(nH/T_P)} \\
&\quad - C\left(2\sin\frac{\pi}{N}\right)^4 \sin\left(\frac{2\pi}{N}n - \theta - \varphi - \frac{4\pi}{N}\right)
\end{aligned} \tag{2.8}
$$

The authors have carried out a detailed analysis of saturation detection using Eqs. (2.5)–(2.8). Here, it has been observed that the accuracy of saturation detection is steadily increased as one moves from 2-point formulas (Eq. 2.5) to 5-point formulas (Eq. 2.8). It is true that any further increase in formulas (beyond 5-point) will definitely reduce the saturation detection error. But at the same time, it will unnecessarily increase the amount of calculation. Hence, in this paper, authors have derived a saturation detection index (D_n) using Eqs. (2.5)–(2.8) and Newton's backward difference formulas [7]. They are given as:

$$D_{n3} = \frac{1}{H}\left[\nabla_n^1 + \frac{\nabla_n^2}{2} + \frac{\nabla_n^3}{3}\right] \quad (2.9)$$

$$D_{n4} = \frac{1}{H}\left[\nabla_n^1 + \frac{\nabla_n^2}{2} + \frac{\nabla_n^3}{3} + \frac{\nabla_n^4}{4}\right] \quad (2.10)$$

where H is the sampling interval.

Taking the difference of Eqs. (2.9) and (2.10), a saturation detection index (D_n) can be calculated and given by Eq. (2.11).

$$D_n = D_{n4} - D_{n3} = \frac{1}{H}\left[0.25i_{2[n]} - i_{2[n-1]} + 1.5i_{2[n-2]} - i_{2[n-3]} + 0.25i_{2[n-4]}\right] \quad (2.11)$$

This index (D_n) is compared with an adaptive threshold to estimate the start and endpoint of CT saturation.

2.1.2 *Condition for CT Saturation Detection*

The value of D_n is much larger than the constant term "$C\left(2\sin\frac{\pi}{N}\right)^4$" available in the sinusoidal part of the Eq. (2.8) during CT saturation. This term is used to derive adaptive threshold (T_h) along with several other terms such as the amount of maximum fault current (I_{max}) estimated using the Fourier algorithm and safety factor (λ) which depends on low pass filter.

Hence, the adaptive threshold is given as below.

$$T_h = \lambda * \sqrt{2} * I_{max} * C * \left(2\sin\frac{\pi}{N}\right)^4 \quad (2.12)$$

The said value of the adaptive threshold is capable to detect small to heavy saturation conditions as it depends on the magnitude of fault current and λ compared to the scheme given in [5] which uses fixed threshold value.

2.2 Proposed Saturation Detection Flowchart

Figure 2.1 shows the flowchart of the proposed algorithm. Initially, current samples of bay CTs are acquired by the data acquisition system through a first-order low pass filter which effectively removes the noise present in the secondary current.

The fault detection algorithm is used to discriminate between the fault and normal conditions [8]. Whenever a fault is detected by the fault detection algorithm, post fault samples of all phases of connected bay CTs are sent to the CT saturation estimation

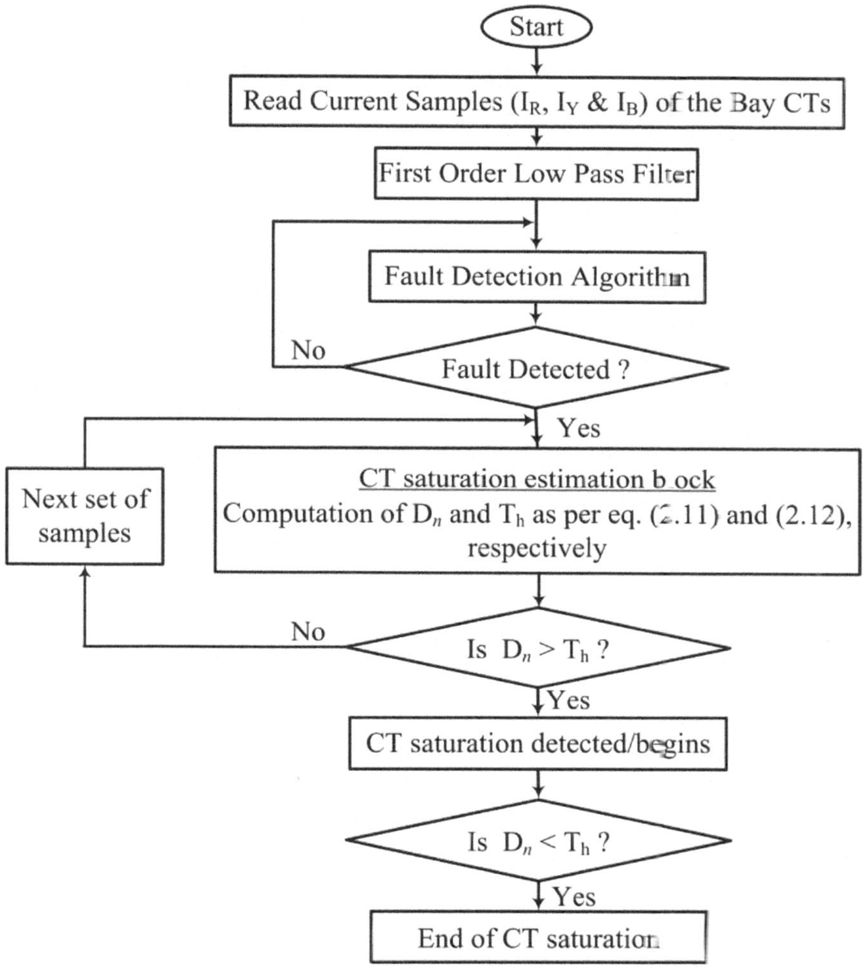

Fig. 2.1 Algorithm of CT saturation detection

block. In this block, the value of D_n is calculated using five-point formulas (Eq. 2.11) for each cycle and is being continuously compared with an adaptive threshold. When the value of D_n exceeds the threshold value, the starting point of CT saturation is detected ($D_n > T_h$), and thereafter end of saturation is noticed when the value of D_n goes below a threshold value.

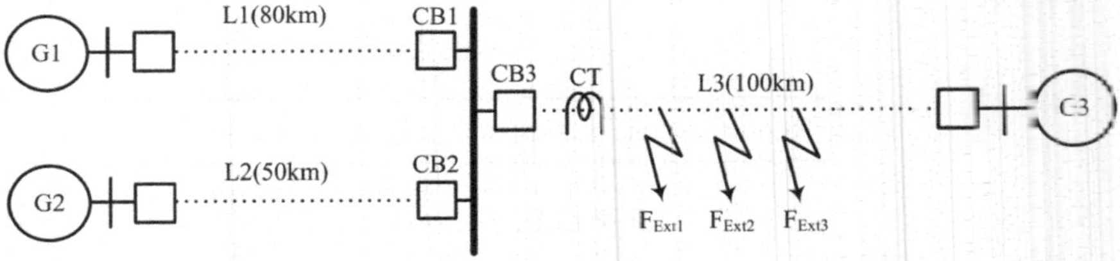

Fig. 2.2 Single line diagram of the power system model

2.3 System Study

Figure 2.2 shows a single line diagram of a power system network consisting of three sources connected to the common bus through bay L1, L2 and L3, respectively. Figure 2.2 is simulated using the PSCAD/EMTDC software packages.

To validate the proposed algorithm, CTs located on bay L3 are analyzed which uses Jiles–Atherton model available in PSCAD/EMTDC software package [9]. All the test cases are generated by simulating faults on bay L3 with varying fault and system parameters. These parameters are Fault Inception Angle (FIA), fault resistance (R_f), types of fault (F_{type}), and Fault Locations (FL) online L3 (F_{ex1}, F_{ex2}, F_{ex3}). The line and source parameters are given in the Appendix. The sampling frequency of [illegible] kHz is used in this study for a system operating at a frequency of 50 Hz.

2.4 Simulation Results and Discussion

The proposed CT saturation detection technique is very fast considering the adaptive threshold. However, just after fault inception, the secondary current has a point of inflection. Hence, D_n may have a large value at the next sample of a fault instant; the proposed algorithm may detect this instant as the start of saturation. To avoid maloperation under this situation, the proposed algorithm starts after a current that exceeds three times the rated secondary current for three successive samples [5].

Different parameters such as remanence flux density, burden resistance, presence of DC offset and noise have been considered for the validation of the proposed scheme. Considering these all parameter values, around 900 simulation cases were generated and the effectiveness of the proposed scheme was verified for all these test cases. However, only a few results are shown in the upcoming section.

2.4.1 *Effect of DC Component and Secondary Burden on CT Saturation*

The performance of the proposed scheme during CT saturation is carried out by simulating different faults on bay L3 at different locations (5, 10, and 20 km) from the bus with varying system parameters. By changing the CT secondary burden resistance, different degrees of saturation can be obtained [10]. Figure 2.3a shows the CT primary and secondary currents and Fig. 2.3b show the value of D_n and threshold (T_h) during R-g fault on bay L3 at 20 km without CT saturation and DC component. It has been observed from Fig. 2.3b that the magnitude of D_n remains well below the adaptive threshold throughout the fault time and hence no saturation detected by the proposed algorithm. Figure 2.3c, d show the performance of the proposed scheme in the presence of decaying DC component and burden resistance ($R_b = 1\ \Omega$). Here, the value of D_n crosses the threshold value (Fig 2.3d) after one cycle elapse from the point of fault inception (start of saturation) and remains above the threshold value for the next three successive cycles. The saturation ends when the value of D_n goes well below the threshold value.

Moreover, Fig. 2.4a–d show the performance of the proposed algorithm for R-Y fault on bay L3 at 5 km during burden resistance (R_b) equals to 3 and 6 Ω, respectively. It is to be noted from Fig. 2.4b, d that the proposed scheme is capable to detect severe CT saturation conditions in the presence of decaying DC component.

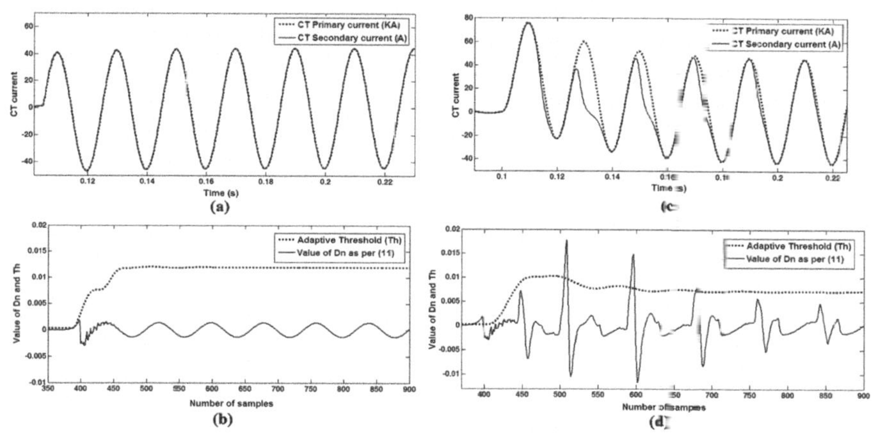

Fig. 2.3 Waveform of CT currents and value of D_n and T_h, **a**, **b** without CT saturation, **c**, **d** with CT saturation

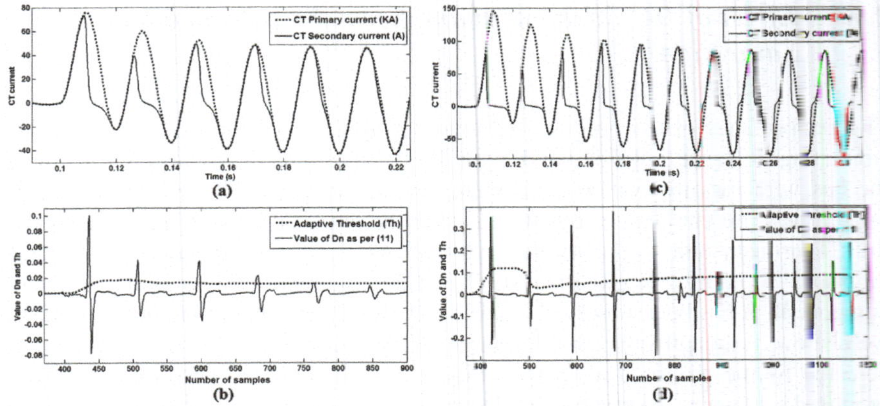

Fig. 2.4 Waveform of CT currents and value of D_n and T_h under CT saturation condition **a**, **b** $R_b = 3\ \Omega$ and **c**, **d** $R_b = 6\ \Omega$

2.4.2 Effect of Remanent Flux on CT Saturation

During the energization of CT in presence of remanent flux in the core, a large part of the secondary current of CT may saturate [2] and residual magnetism may reach up to 90% of the saturation flux [11]. Figure 2.5a–d show the CT currents and value of D_n and T_h, respectively, for a three-phase (R-Y-B) fault at 10 km on bay L3 with 0.5 Ω burden resistance during 0% and 90% remanent flux density (set in the core of CT prior to the inception of fault). It is to be noted from Fig. 2.5b, d that the proposed algorithm is capable to detect the saturation interval (by comparing the value of D_n and threshold) irrespective of the level of remanence flux previously present in the core of CT.

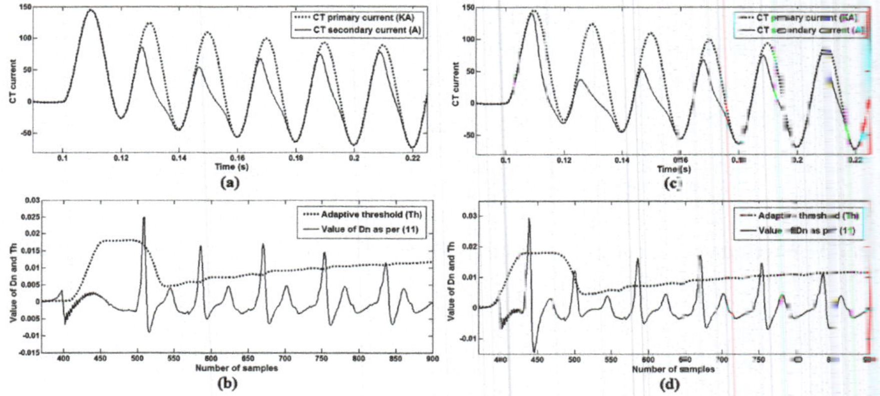

Fig. 2.5 Waveform of CT currents and value of D_n and T_h during **a**, **b** 0% remanence flux and **c**, **d** 90% remanence flux

2.4.3 *Effect of Noise Superimposed in Secondary Current*

Acquired current signals from PSCAD software are polluted with white Gaussian noise by considering different signal-to-noise ratios (SNR) in the MATLAB environment. The SNRs are set to 20, 30, and 40 db to pollute the original current signals. Thereafter, to diminish the higher-order harmonics and noise, a low pass first-order Butterworth filter is used. With a sampling frequency of 4 kHz, the cut-off frequency of the filter is gradually decreased from 1600 to 200 Hz for perfect saturation detection. Figure 2.6a shows the CT currents and (b) value of D_n & threshold during R-Y-g fault on bay L3 at 5 km with $R_b = 3\ \Omega$, SNR = 40 db, and cut-off frequency = 300 Hz. It has been observed form Fig. 2.6b that the proposed algorithm accurately detects the start and end of saturation. It is to be noted that at low cut-off frequency, the proposed algorithm gives more efficient results in terms of saturation detection in the presence of harmonics and noise.

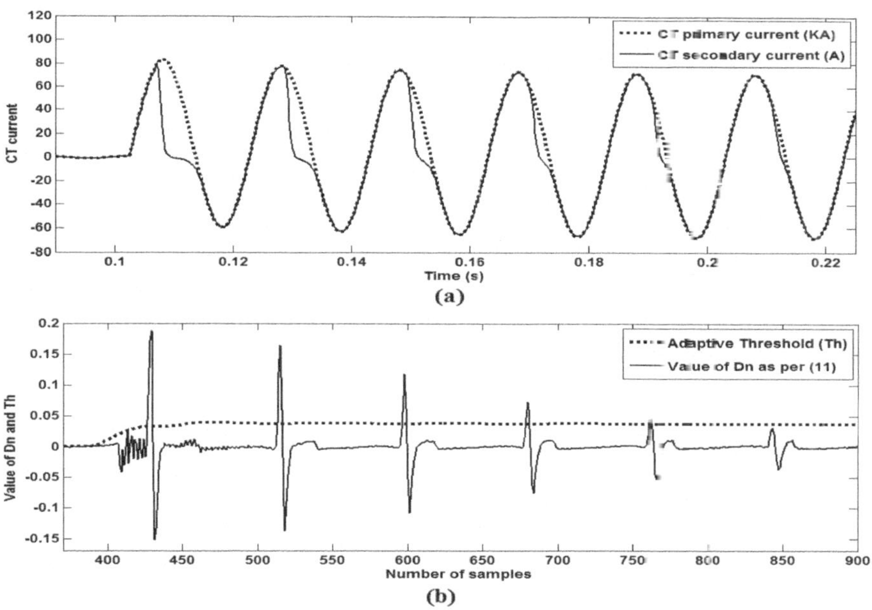

Fig. 2.6 Waveform of CT primary and secondary current **a** and value of D_n and T_h, **b** during SNR = 40 db contained by CT secondary signals

2.4.4 Effect of Types of Fault and Fault Inception Angle (FIA)

The system shown in Fig. 2.2 was subjected to various types of faults. The results presented in Figs. 2.3, 2.4, 2.5 and 2.6 demonstrate that the proposed algorithm detects CT saturation condition for both symmetrical and asymmetrical faults.

Moreover, various simulation cases have been generated by varying the FIA between 0° and 360° to identify its effect on CT saturation. Figure 2.7a, b show the CT currents and value of D_n & T_h, for R-g fault applied at 5 km on bay L3 with $R_b = 3\ \Omega$ and FIA $\theta = 45°$. The simulation results for the same fault condition with FIA $\theta = 135°$ and $R_b = 5\ \Omega$ are shown in Fig. 2.7c, d.

It has been observed from Fig. 2.7b, d that though the magnitude of decaying DC component is affected by FIA, the proposed scheme correctly identifies the start and endpoints of CT saturation.

2.5 Proposed Compensating Algorithm

An efficient CT compensation algorithm can significantly reduce errors in measured current signals during saturation of CT. In this paper, the unsaturated portion of the secondary current signal (as detected by Newton's backward difference formulas) is used to compensate for the saturated portion. Moreover, a Modified Discrete Fourier Transforms (MDFT) algorithm [12] is used with a short moving data window to estimate the phasor parameters of an unsaturated section of CT secondary current. The MDFT filter accurately estimates both phasor magnitude and phase angle by utilizing 12 samples of unsaturated portions of the current signal (3 ms) by eliminating

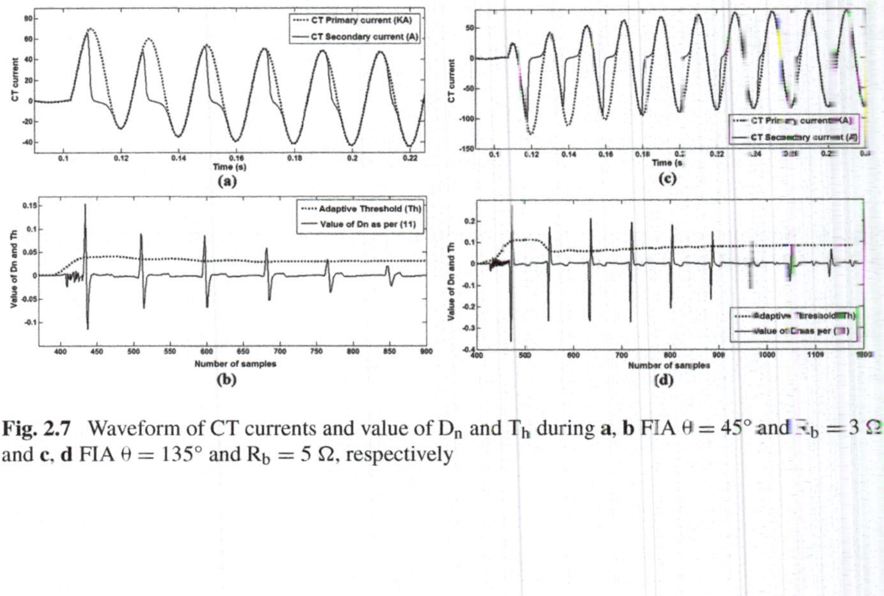

Fig. 2.7 Waveform of CT currents and value of D_n and T_h during **a**, **b** FIA $\theta = 45°$ and $R_b = 3\ \Omega$ and **c**, **d** FIA $\theta = 135°$ and $R_b = 5\ \Omega$, respectively

integer harmonics and decaying DC components [12]. The sampling frequency of the proposed compensating algorithm is the same as that of the saturation detection algorithm (4 kHz).

The proposed compensating algorithm has been validated on various saturated CT secondary current signals. However, one sample case is explained here by considering the current signal of Fig. 2.7c. The first window of Fig. 2.8a shows sampled one cycle fundamental frequency component of Fig. 2.7c during the saturation of CT (550–630 samples). It has been observed from the second window of Fig. 2.8a that the calculated fault current magnitudes with MDFT filter are imperfect and unstable during the saturation period (561–578 samples and 604–618 samples) whereas, it shows stable magnitude during an unsaturated portion (580–602 samples) of the current waveform. During the sinusoidal portion of the current signal, the average value of D_n remains almost close to zero (Fig. 2.8a). Hence, based on the proposed saturation detection algorithm, when $D_n \cong 0$, an unsaturated portion (T_{UNSAT}) is distinguished from a saturated portion (T_{SAT}) and T_{UNSAT} is precisely estimated using short moving window length (N = 12 samples) by MDFT filter. In the proposed method, the estimated current magnitude (M_{UNSAT}) and phase angle (θ_{UNSAT}) for the duration of T_{UNSAT} has been utilized for the correction of the saturated current signal.

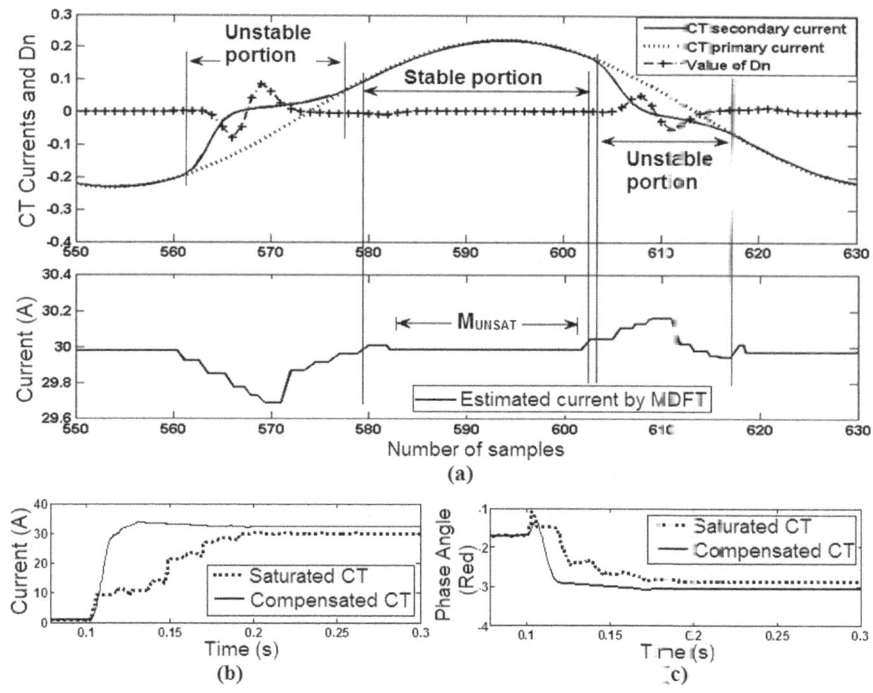

Fig. 2.8 **a** CT currents and estimated current magnitude by MDFT filter, **b** compensated current magnitude, and **c** compensated phase angle of the CT for the current signal of Fig. 4c

During the operation of the compensation algorithm, the estimated parameters i.e. M_{UNSAT} and θ_{UNSAT} obtained from MDFT are directly allocated to the output of the algorithm. On the other hand, during the saturated sampled portion (T_{SAT}), the estimated M_{UNSAT} and θ_{UNSAT} are kept unchanged from the value of the last T_{UNSAT} segment. Thus, the algorithm prevents the inaccurate estimation of phasor parameters during the detected saturation interval and seizes unsaturated phasor parameters (values of last T_{UNSAT} segment) once the transition time of the MDFT filter expires. Figure 2.8b, c show the calculated values of magnitude and phase angle, respectively, with the saturated condition of CT and after its compensation for the current signal of Fig. 2.8c. It has been observed from various simulation results that the proposed compensation algorithm accurately reconstructs the distorted portion of the current signal and provides effective output.

2.6 Practical Validation of the Proposed Algorithm

2.6.1 Hardware Setup

In order to evaluate the performance of the proposed algorithm during CT satura-tion condition, a laboratory test bench, as shown in Fig. 2.9, is developed. Here, protective class (5P10) CT having ratio = 10/5 A, burden = 5 VA and voltage rating = 660 V is used. Further, the relay testing kit is used to inject high current (0–250 A) in the primary of CT and the variable rheostat is used as a secondary burden resistance. In order to record the waveform of CT secondary current, a high-resolution Digital Storage Oscilloscope (DSO) along with a clamp-on type current sensor probe is also used. A sampling of the recorded current signal is carried out at a rate of 80 samples/cycle in DSO. Subsequently, these sampled data are migrated in MATLAB software using the USB port of DSO and further utilized for testing of the proposed CT saturation algorithm.

2.6.2 Results of Prototype

In order to validate the proposed algorithm, various cases have been generated by changing burden resistance from 0 to 12 Ω and primary current of CT from 10 to 120 A. Figure 2.10a shows the CT secondary current during saturation along with a zoomed view of a certain portion of the signal, captured by DSO, during 100 A primary current and $R_b = 12\ \Omega$. The performance of the proposed algorithm in terms of D_n and T_h are shown in Fig. 2.10b.

It has been observed from Fig. 2.10b that the proposed scheme correctly detects severe CT saturation condition as the value of the detection index exceeds a threshold value (detects only starting point as there is no endpoint for the collected data).

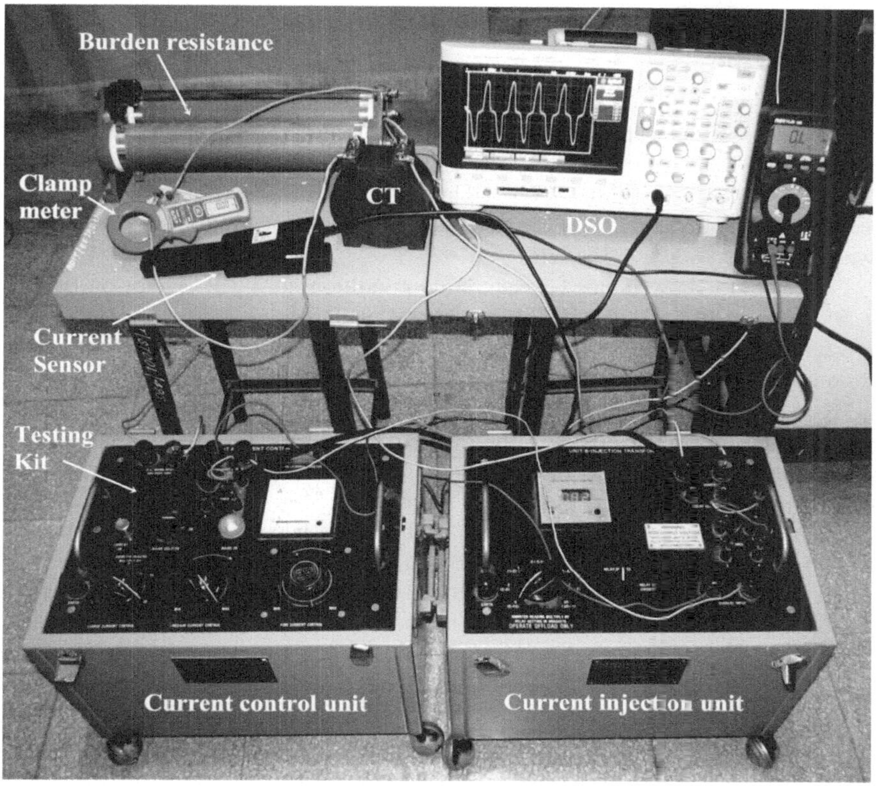

Fig. 2.9 Hardware setup of laboratory test bench

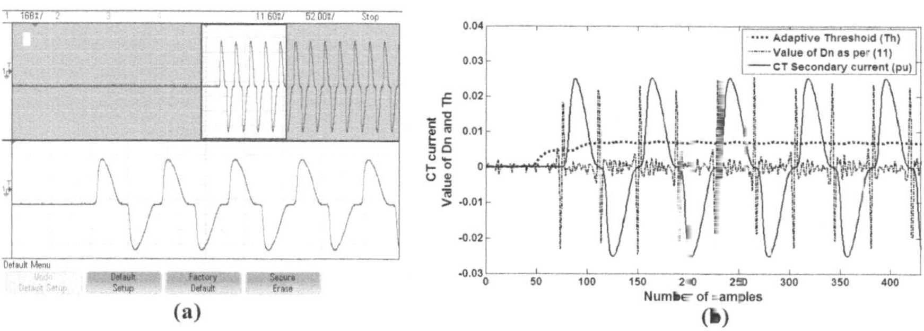

Fig. 2.10 **a** CT secondary current captured by DSO and **b** values of D_n and T_h for the said condition

2.7 Comparison of the Proposed Algorithm with Existing Scheme

It has been observed that the schemes based on second and third difference functions [5, 6] may not be able to identify the endpoint of saturation and may operate in case of very low saturation of CT. Conversely, the proposed algorithm provides accurate results irrespective of the level of saturation compare to above mentioned two schemes as shown in Fig. 2.11.

Figure 2.11a show minor CT saturation condition during B-g fault on bay 13 at 50 km with $R_b = 0.06\ \Omega$. The magnitude of derivative Del_2, Del_3 and D_n & threshold T_{h1}, T_{h2}, and T_h during the second difference, a third difference of the sampled currents, and using five-point formulas of the proposed algorithm is shown in Fig. 2.11b–d, respectively. It is to be noted from Fig. 2.11b, c that the value of Del_2 and Del_3 remains well below the respective threshold T_{h1} and T_{h2} under minor CT saturation condition. On the other hand, as shown in Fig. 2.11d, the proposed algorithm accurately detects the saturation interval.

Further, to compare the performance of the proposed scheme with the Wavelet-based technique [13], another test case is generated and results are presented in Fig. 2.12. It is to be noted from Fig. 2.12b, c that the magnitude of detailed coefficient 'd1' obtained from Daubechies-4 (db4) mother wavelet analysis is quite lower than the value of D_n given by the proposed scheme. Hence, the proposed scheme gives satisfactory during minor CT saturation conditions compare to the wavelet technique.

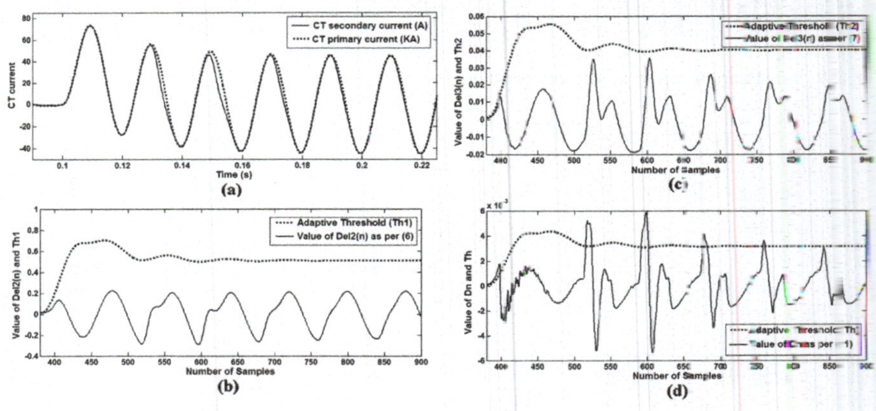

Fig. 2.11 **a** CT currents, **b** value of del_2 and T_{h1} during second difference, **c** value of del_3 and T_{h2} during third difference, **d** value of D_n and T_h of the proposed algorithm

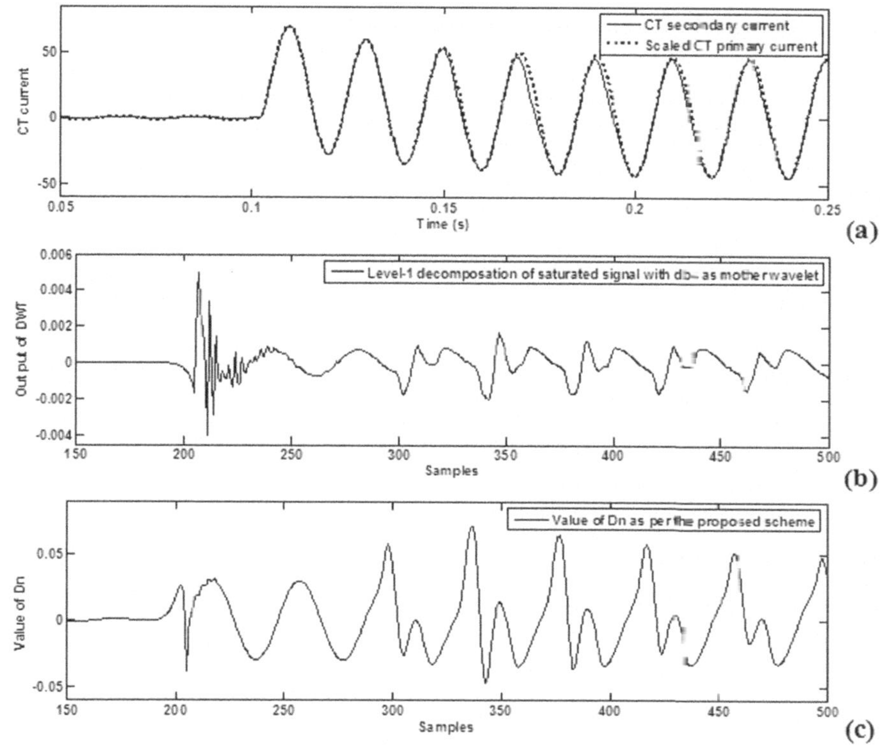

Fig. 2.12 **a** CT currents, **b** output of wavelet technique and **c** value of D_n

2.8 Summary

This fraction presents a new algorithm for the detection and compensation of CT saturation conditions. The algorithm is based on a saturation detection index which is obtained using five-point Newton's backward difference formulas. Validation of the proposed algorithm is carried out using a CT model available in PSCAD/EMDC software by considering parameters such as remanence flux, FIA, burden resistance, and presence of DC offset and noise. A compensating algorithm has been proposed which effectively reconstructs the saturated portion of CT secondary signals. The proposed algorithm is also validated using various CT saturation cases generated in the laboratory environment. Also, based on the comparative evaluation, the performance of the proposed scheme is found to be superior compare to the existing schemes. Moreover, results obtained from the analysis demonstrate the effectiveness of the proposed algorithm for accurate detection and compensation of CT saturation condition.

2.9 Published Article Based on This Work

N. G. Chothani and B. R. Bhalja, "New Algorithm for current transformer saturation detection and compensation based on derivatives of secondary currents and Newton's backward difference formulae," IET Gener. Transm. Distrib., vol. 8, no. 5, pp. 841–850, May 2014.

Appendix

Source Data

Positive-sequence and Zero-sequence impedance of G1, G2 and G3 = 1.5 − j8.2 Ω, 0.0174 + j0.199 Ω, 1.307 + j14.942 Ω and 0.035 + j0.098 Ω, 0.00435 + j0.0498 Ω and 0.817 + j9.961 Ω, respectively.

Load angle of G3 is set at −5°, Frequency = 50 Hz, Voltage = 220 kV.

Transmission-line Data

Positive and Zero-sequence impedance = 0.0297 + j0.332 Ω/km and 0. 62 − j1.24 Ω/km.

Positive and Zero-sequence capacitance = 9.23 nF/km and 6.72 nF/km.

CT Data

CT ratio: 1500/5 Amp,

Secondary winding resistance and inductance = 0.5 Ω and $0.8e^{-3}$ H, respectively.

References

1. Bhalja B, Maheshwari RP, Chothani NG (2017) Protection and switchgear, 2nd edn. Oxford University Press, New Delhi, India
2. WSC Council, Relaying current transformer application guide. Relay work group
3. PSCAD Research Center (2005) EMTDC-transient analysis for PSCAD power system simulation. Winnipeg, MB, Canada
4. Phadke AG, Thorp JS (2009) Computer relaying for power systems. Wiley
5. Kang YC, Ok SH, Kang SH (2004) A CT saturation detection algorithm. IEEE Trans Power Deliv 19(1):78–85
6. Dashti H, Sanaye-Pasand M, Davarpanah M (2009) Fast and reliable CT saturation detection using a combined method. IEEE Trans Power Deliv 24(3):1037–1044
7. Goyal M (2007) Computer-based numerical & statistical techniques. Infinity Science Press LLC, Hingham, Massachusetts, New Delhi, India
8. Mohanty SR, Pradhan AK, Routray A (2008) A cumulative sum-based fault detector for power system relaying application. IEEE Trans Power Deliv 23(1):79–86
9. EU Manual (2005) Manitoba HVDC Research Center. Winnipeg, MB, Canada

10. Annakkage UD, McLaren PG, Dirks E, Jayasinghe RP, Parker AD (2000) A current transformer model based on the Jiles-Atherton theory of ferromagnetic hysteresis. IEEE Trans Power Deliv 15:57–61
11. IEEE Guide for Protective Relay Applications to Power System Buses. IEEE Std C37.234-2009, Nov 2009, pp C1-115
12. Yu S-L, Gu J-C (2001) Removal of decaying DC in current and voltage signals using a modified Fourier filter algorithm. IEEE Trans Power Deliv 16(3):372–379
13. Hong YY, Chang-Chian PC (2008) Detection and correction of distorted current transformer current using wavelet transform and artificial intelligence. IET Gener Transm Distrib 2(4):566–575

Chapter 3
Phasor Angle Based Differential Protection of Power Transformer

Conventional protection schemes of a power transformer may operate during abnormalities such as Inrush condition, CT saturation during an external fault, and high resistance internal fault conditions. This chapter presents inrush detection based on the average angle of 2nd order derivative of differential current. The magnitude and phase angle of primary and secondary currents are estimated using the Modified Full Cycle DFT algorithm. Normally, during an internal fault, the differential current is well above the restraining current; however, the same condition arises during external fault with heavy CT saturation. Hence, the phase angle comparison based scheme is combined with a percentage of biased differential protection. During an internal fault, the phase angle difference between primary and secondary current is less than 90°, whereas it differs more than 90° for any external fault conditions. In the proposed scheme, 315 MVA, 400/220 kV transformer is considered for validation of various fault conditions simulated in PSCAD™ software. Moreover, the proposed scheme is validated on a developed laboratory prototype of 2 kVA, 230/115 V single phase transformer using an ATmega328 microcontroller. The proposed algorithm is effectively validated on both, simulation and hardware to discriminate inrush, internal and external fault by considering various systems and fault parameters.

3.1 Literature Review

As very expensive and sensitive equipment in the power system, the transformer needs an accurate protective scheme against various abnormal situations. Core saturation is one most tough issues due to the generation of inrush current and causes maloperation in transformer protection. Recently, cold-rolled grain-oriented (CRGO) steel is used for the manufacturing of the core of the power transformer. The transformer manufacturing technology should be such that the saturation of core and noise during its operating life may minimum. This can be achieved by amorphous core material, asymmetric core assembly without affecting magnetic properties, step

D. Patel and N. Chothani, *Digital Protective Schemes for Power Transformer*, Power Systems, https://doi.org/10.1007/978-981-15-6763-6_3

lap joints, and proper sizing of the core. Residual flux plays a major role in the saturation of the magnetic core. For a residual flux density maximum inrush current (become three times) is drawn when the transformer is switched on at the instant when the applied voltage is zero [1]. It can be observed that the current waveform is completely offset in the first few cycles with wiping out of alternate half-cycles because the flux density is below saturation value for these half-cycles. This current waveform is containing higher-order harmonics. Hong-ming et al. [2] explained the relation and effect between magnetizing inrush and sympathetic inrush based on 2nd harmonic component and changing the neutral grounding mode. Hunt et al. [3] elaborate on the disadvantage of 2nd harmonic based technique. Many researchers have applied current waveform characteristic based analysis. Hong et al. [4] presented problem definition Waveform Complexity Analysis (Fractal Analysis) of differential current to define inrush in power transformer. Even though, an internal fault within the inrush condition case is not tested.

In the past, many researchers have proposed transformer protection schemes based on a mathematical model, classifier, and decomposing techniques. Barbosa et al. [5] elaborated fuzzy and Clark's transform for transformer protection. However, limited test cases were validated with a higher time of operation. Tripathy et al. [6] proposed an optimal probabilistic neural network for the unit type of transformer protection. Mohamed et al. [7] described neural network-based fault discrimination in power transformer. de Faria et al. [8] demonstrated ANN with Bayesian Classifier (BC) and swarm optimization method to discriminate fault and abnormal conditions in the transformer. Guzman et al. [9] evaluated the zero sequences component-based transformer protective scheme with an unsupervised artificial neural network. Moravej et al. [10] detailed S-transform for power transformer protection. To specify the feature, Probabilistic Neural Network (PNN) and the Support Vector Machine (SVM) are involved. Conversely, the training and testing of data in the techniques mentioned are complex and time-consuming process; this will delay the trip signal. Moreover, there are no exact rules for setting the number of hidden layers, the number of layers, and the type of transfer function in NN techniques.

Stanbury and Djekic [11] explained the effect of CT saturation on transformer protection. Ahmadi et al. [12] suggested discrimination of inrush current with internal fault using a sine-wave least-squares curve fitting method. However, various fault conditions with high resistance, fault location on transformer winding, fault inception angle, etc. is not tested in this scheme. Khan and Sidhu [13] demonstrated transformer differential protections based on the directional comparison technique, on the other hand, the cost of voltage transformer increases. Valsan and Swarup [14] elaborated transformer protection with high-frequency power directional signals based on Wavelet, so far the number of tested data is too less. Jettanasen and Ngaopitakkul [15] used Discrete Wavelet Transform (DWT) as a spectrum analysis to discriminate fault in transformer protection. However, the scheme has less fault classification efficiency. Naumov and Shevtsov [16] explained the mathematical modeling of the current transformer for differential protection of the transformer. Admittance, impedances are also affected during abnormal conditions in the power system. On positive sequence admittance based transformer, protection is implemented by Eissa et al. [17] on the prototype. However, load variation conditions and fault cases with various switching angles are not considered. Rahmati et al. [18] described multi-criteria decision-making based power transformer protection however, the operating time of the scheme is high. Noshad et al. [19] presented Clarke's Transform and Discrete Wavelet Transform to mitigate mal-operation of transformer differential protection due to a high CT saturation phenomenon. However, these schemes were validated for a limited number of test cases and also require high computational time. Although, results presented in the aforesaid techniques are encouraging in such a particular trend realization as a practical approach is challenging due to complicated algorithms.

Discrimination of core saturation based on current harmonic techniques is old and unreliable. This piece of writing presents inrush detection based on the angle of second-order derivative-based differential current and discrimination of internal and external fault based on percentage biased differential scheme with a combination of phase angle comparison technique for power transformer protection. The proposed algorithm works accurately during internal fault and remains stable against all abnormal conditions like inrush, external fault, CT saturation during an external fault, and overloading conditions. To validate the algorithm, various test cases are performed on PSCAD™ software [20] as well as on the prototype model developed in the laboratory. Section 3.2 includes problem definitions and potential solutions with the proposed algorithm with system modeling. Section 3.3 elaborates on the simulation results. Sections 3.4 and 3.5 covers prototype development and practical validation of the proposed schemes. Section 3.6 shows novelty of the algorithm.

3.2 A Proposed Transformer Protection Technique

3.2.1 Problem Description and Solution

The percentage biased differential principle is widely used in the field for a few decades and performing very well. However, only the current magnitude based technique may operate during inrush and external fault with heavy CT saturation. This is due to the disparity in primary and secondary currents of the transformer at a relaying point during an external fault. Hence, to overcome the said problem, angle of second-order derivative differential current for inrush detection and combined percentage biased differential and phase angle difference based technique is proposed to discriminate external fault with heavy CT saturation. Figure 3.1 demonstrates the phase angle difference of primary (I_1) and secondary (I_2) current during an internal and external fault condition. It is to be noted from Fig. 3.1 that the phase angle between primary and secondary side of a transformer during internal fault condition will be minimum (Ideally 0°), on the other hand, the phase angle difference will be approximately 180° under normal and external fault conditions. However, due to the heavy saturation of CT during an external fault, the magnitude-based scheme (biased differential)

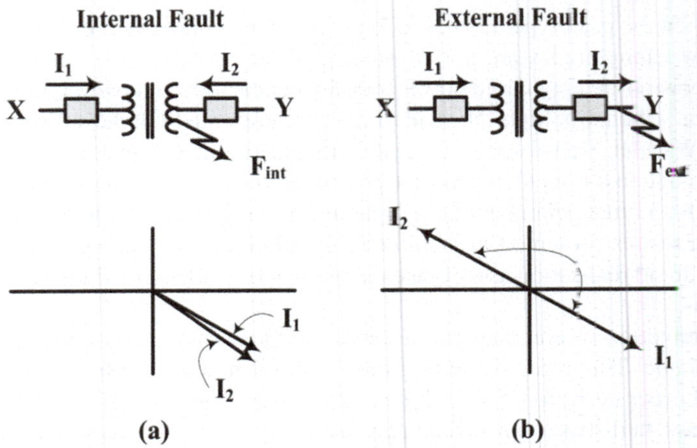

Fig. 3.1 Phasors of primary and secondary current during **a** Internal and **b** External fault

fails and may operate. Conversely, during the same situation, the phase angle difference may retard from 180° but never set below 90°. Moreover, during high resistance internal fault, the phase angle difference may enhance from 0° but never set above 90° [21]. This technique is authenticated in the proposed work to detect all transformer internal faults accurately and discriminate all external abnormalities.

Figure 3.1a shows the phasor of primary current (I_1) and secondary current (I_2) fall in the same quadrant (less than 90°) during an internal fault. Whereas they are almost out of phase and fall in different quadrants (greater than 90°) during external fault conditions. The threshold setting of angle difference may vary depending on the connection of the transformer (star-star or delta-star).

3.2.2 *Proposed Algorithm*

Figure 3.2 describes the proposed algorithm for discrimination of inrush, internal and external fault on the transformer.

The primary and secondary currents of CTs are captured through a data acquisition system. The sampling frequency of 4 kHz (80 samples/cycle) at an operating frequency of 50 Hz is used in this algorithm. After that, the current samples are migrated from PSACD™ software, magnitude, and phasor estimation with MATLAB programming of Modified Full Cycle Discrete Fourier Transform (MFCDFT) [22].

The full cycle MDFT algorithm which can extract fundamental frequency components from a given input signal is presented in this work. Moreover, this is used to remove the DC component and harmonics when applied in the filter algorithm of the digital relay. Consider a full cycle time period T and continuous sinusoidal

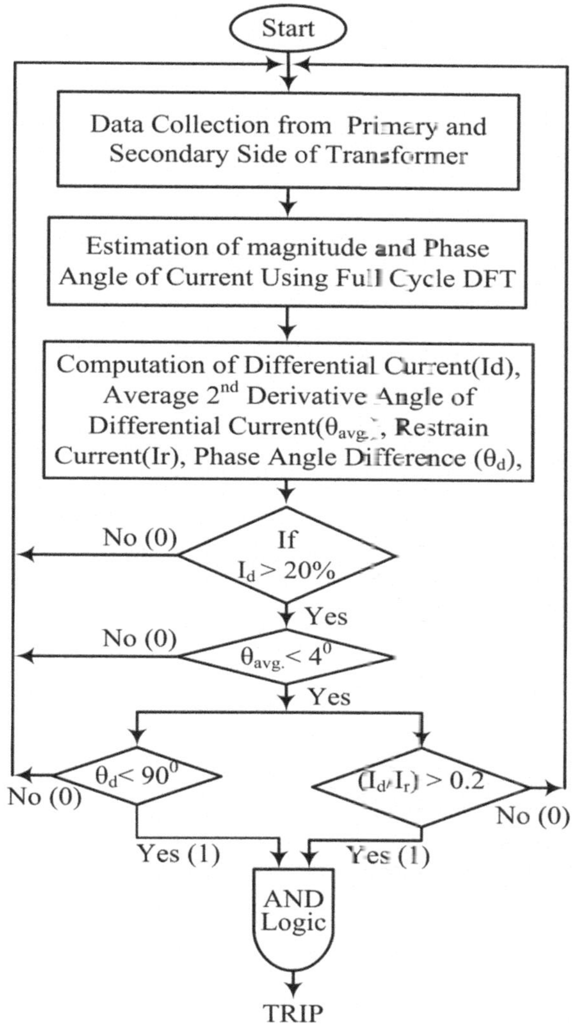

Fig. 3.2 Proposed algorithms for transformer protection

(voltage or current) signal $f(t)$ which contains the DC component and N-2 order harmonics. If the sampling frequency is considered as f_S then N is the sampling rate for a fundamental frequency period. The sample period/time step of the algorithm is $\Delta T = T/N$. Then, $f(t)$ and the Kth sample signals $f(k)$ are represented by Eqs. 3.1 and 3.2.

$$f(t) = A_0 + \sum_{n=1}^{N-2} A_n \cos(n\omega t + \theta_n) \tag{3.1}$$

$$f(k) = A_0 + \sum_{n=1}^{N-2} A_n \cos\left(\frac{2nk\pi}{N} + \theta_n\right) \tag{3.2}$$

Fundamental frequency complex phasor contains both, the real part $F_{r(k)}$ and imaginary part $F_{i(k)}$.

$$F_{r(k)} = \tfrac{2}{N} \sum_{r=k-N+1}^{k} * f(r) * \cos\left(\frac{2r\pi}{N}\right) \tag{3.3}$$

$$F_{i(k)} = \tfrac{-2}{N} \sum_{r=k-N+1}^{k} * f(r) * \sin\left(\frac{2r\pi}{N}\right) \tag{3.4}$$

Analog low pass filter can remove higher-order harmonics with relative easiness and, simultaneously, produces the new decaying dc time constant. Fortunately, the new time constant is known and is obtained according to the characteristic equation of a low pass filter. Here, we discuss modified FCDFT using first-order low pass filter. Where $\xi(t)$ denotes a voltage or current signal before low pass filter during the fault period and τ represents the decaying dc time constant. The time constant of low pass filter is τ_1.

$$\xi(t) = A_0 + \sum_{n=1}^{\infty} \mathrm{A_n cos(n\omega t} + \theta_\mathrm{n}) + \mathrm{Be}^{-t/\tau} \tag{3.5}$$

First order low pass filter characteristics:

Fundamental frequency amplitude gain: K_{A1}, Fundamental frequency phasor angle shift: $K_{\theta 1}$.

The output signal of the first order low pass filter is $f(t)$.

$$f(t) = A_0 + \sum_{n=1}^{N-2} \mathrm{C_n cos(n\omega t} + \phi_\mathrm{n}) + \mathrm{De}^{-t/\tau} + \mathrm{D_1 e}^{-t/\tau_1} \tag{3.6}$$

Using FCDFT, it gives,

$$F_{r(N)} = \frac{2}{N} \sum_{k=1}^{N} * f(k) * \cos\left(\frac{2r\pi}{N}\right)$$

$$F_{r(N)} = C_1 \cos\phi_1 + \frac{2}{N} \sum_{k=1}^{N} \left[\mathrm{De}^{-k\Delta T/\tau} + \mathrm{D_1 e}^{-k\Delta T/\tau_1}\right] * \cos\left(\frac{2k\pi}{\mathrm{N}}\right) \tag{3.7}$$

$$F_{i(N)} = -\frac{2}{N} \sum_{k=1}^{N} * f(k) * \sin\left(\frac{2r\pi}{N}\right)$$

$$F_{i(N)} = C_1 \sin \phi_1 - \frac{2}{N}\sum_{k=1}^{N}\left[\mathrm{De}^{-\mathrm{k}\Delta T/\tau} + \mathrm{D}_1\mathrm{e}^{-\mathrm{k}\Delta T/\tau_1}\right] * \sin\left(\frac{2\mathrm{k}\pi}{\mathrm{N}}\right) \quad (3.8)$$

Let, $R = e^{-\Delta T/\tau}$ (unknown), $S = e^{-\Delta T/\tau_1}$ (known)

$$\frac{N\left(F_{r(N+1)} - F_{r(N)}\right)}{2\cos(2\pi/N)} = D * R\left(R^N - 1\right) + D_1 * S\left(S^N - 1\right) \quad (3.9)$$

$$\frac{N\left(F_{r(N+2)} - F_{r(N+1)}\right)}{2\cos(4\pi/N)} = D * R^2\left(R^N - 1\right) + D_1 * S^2\left(S^N - 1\right) \quad (3.10)$$

$$\frac{N\left(F_{r(N+3)} - F_{r(N+2)}\right)}{2\cos(6\pi/N)} = D * R^3\left(R^N - 1\right) + D_1 * S^3\left(S^N - 1\right) \quad (3.11)$$

$$(3.10) - (3.9) \times S = D * R\left(R^N - 1\right)(R - S) \quad (3.12)$$

$$(3.11) - (3.10) \times S = D * R^2\left(R^N - 1\right)(R - S) \quad (3.13)$$

Hence, dividing Eq. 3.12 by Eqs. 3.13, (3.13)/(3.12) we get *R*. Using *R*, *S*, and (3.12) obtain *D*, then using *R*, *D*, and (3.9) obtain D_1. Using *R*, *D*, D_1 and (3.7) obtain $C_1 \cos \phi_1$. Also, using *R*, *D*, D_1 and (3.6) obtain $C_1 \sin \phi_1$.

Thus, $A_1 = C_1/K_{A1}$ and $\theta_1 = \phi_1 + K_{\theta 1}$

Equations 3.3 and 3.4 represent FCDFT. Hence, when $\mathrm{K} \geq \mathrm{N}$ following equations are obtained:

$$F_{r(k)} = A_1 \cos\theta_1 \quad (3.14)$$

$$F_{i(k)} = A_1 \sin\theta_1 \quad (3.15)$$

Amplitude,

$$A_1 = \sqrt{F_{r(k)}^2 + F_{i(k)}^2} \quad (3.16)$$

Phase angle,

$$\theta_1 = \tan^{-1}(\mathrm{F}_{\mathrm{i(k)}}/\mathrm{F}_{\mathrm{r(k)}}) \quad (3.17)$$

Thus algorithm calculates the phasor value of primary and secondary currents. Based on this, computation of differential current, average angle of 2nd order derivative of differential current are calculated as below.

$$DifferentialCurrent = I_d = |I_1 + I_2| \quad (3.18)$$

Second-order Derivative of Differential current,

$$\Delta = \left(\frac{d^2 I_D}{dt^2}\right) \tag{3.19}$$

Angle of the second-order derivative of differential current,

$$\theta = \arctan(\Delta) Degree \tag{3.20}$$

Average of the calculated angle (θ),

$$\theta_{\text{arg}} = \frac{1}{m_i - n_i} \int_{m_i}^{n_i} \theta(t) dt \tag{3.21}$$

$[m_i, n_i]$ are the different time interval over which the average value of θ is estimated.

The second-order derivative (Δ) of the estimated differential current provides useful information about the existence of fault and inrush. Moreover, it is to be proposed that the arctangent of the second derivative of differential current discriminates against the inrush from internal fault. Equations 3.18 to 3.21 describes the process of calculation of the average angle (θ_{avg}) for the proposed algorithm. It is to be stated that the calculation is done on every cycle bases in form of sliding window mode, hence, any further disturbance generated within the first disturbance again one post-disturbance cycle is considered in the estimation of average angle (θ_{avg}). It is observed that the value of θ_{avg} will be approximately 1°–4° in case of internal fault and it will be always more than 4° during magnetizing inrush condition. Only under the symmetrical nature of waveform for internal fault θ_{avg} is up to zero degrees but even under the unsymmetrical waveform of an internal fault, it varies maximum up to 3°. Hence, the threshold limit of 4° is considered due to the mismatch of CT and transient decaying DC components to distinguish internal fault from inrush conditions. Hence, 4 degrees as a cut off is taken by considering the sensitivity of the relay for the detection of inrush in transformer winding against fault condition. Once the inrush situation is alienated, the algorithm returns to the next data sample collection. On the other hand, if a fault condition is detected, the discrimination of internal faults with all other external abnormalities is carried out by phasor angle comparison and biased differential principle.

As shown in Fig. 3.1, during external faults (F_{ext}) ignoring the load current, we get CT secondary current so that, $I_1 = -I_2$ and thus: $|I_d| \cong 0$ and $|I_r| \gg |I_d|$, on the other hand, during internal fault (F_{int}), I_1 and I_2 are almost in phase and therefore $|I_d| \gg |I_r|$. However, this magnitude based scheme alone may operate in case of an external fault with heavy CT saturation. Thus, the phase angle comparison (θ_d) based technique is combined with the biased differential principle for transformer protection.

The restraining current I_r and phase angle difference (θ_d) between primary and secondary currents are estimated in the next step as Eqs. 3.22 and 3.23.

$$Restraining\ Current,\ I_r = \frac{1}{2}|I_1 - I_2| \tag{3.22}$$

$$Phasor\ Angle = \theta_d = \left|\overrightarrow{\theta_1} - \overrightarrow{\theta_2}\right| \tag{3.23}$$

If the differential current is greater than 20 percent of restraining current (computation based on system parameters) [23, 24] and simultaneously, primary and secondary currents phase angle difference is smaller than 90° [21] than trip signal will be issued (internal fault) else otherwise, it is blocked (external fault). Due to less mathematical computations proposed scheme performed high-speed discrimination presently even on hardware in digital relays.

3.2.3 System Modeling

To validate the proposed scheme, authors have simulated the power system as shown in Fig. 3.3 in PSCAD™ software. A three-phase power transformer with variable tapping on each winding is developed in PSCAD to simulate internal fault at different locations. A Thevenin's equivalent generator is connected to 315 MVA, 400/220 kV, Y-Y power transformer through the 400 kV transmission line, and 220 kV side of a transformer are connected to the infinite bus. The system parameter is given in Appendix.

Various internal faults are accounted for the different percentages of winding from a terminal of the transformer including terminal faults considering fault resistance. Moreover, various CT saturation conditions are also simulated for internal as well as external fault conditions to test the proposed algorithm. Furthermore, the proposed technique is validated for different types of faults at different locations (transformer winding and transmission line) considering the fault inception angle (FIA) and power flow angle. Table 3.1 shows the variation in system and fault parameter to generate 1080 internal faults and 1440 external faults data to test the algorithm.

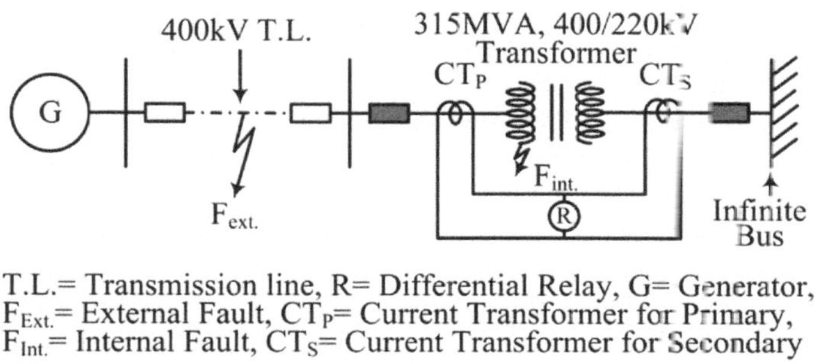

Fig. 3.3 Circuit diagram of power system

Table 3.1 Various fault and system parameter values considered

Fault cases	F_L (% of transformer winding/line)	R_f (Ω)	Fault type (F_{type})	FIA (deg.)	Load angle δ (deg.)
Internal fault in winding (1080)	0, 25 and 50% of winding (primary side) Three values (3)	0, 5 and 10 Three values (3)	L-g (3 No.) L-L (3 No.) L-L-g (3 No.) L-L-L-g (1 No.) Ten types of Fault (10)	0°, 25°, 45° and 90° Four values (4)	0°, 5° and 10° Three values (3)
External to transformer (1440)	5, 25 and 50% of 400 kV line (3) and on 220 kV bus (1) Four values (4)	0, 10 & 20 Three values (3)			

3.3 Simulation Results with Discussion

For testing of the proposed algorithm, various test cases are simulated on considered power system (Fig. 3.3). Different types of fault and system parameters considered during the simulation for validation are as per Table 3.1. Among all the test cases, internal fault and external faults are applied at 0.2 s, and post-fault data are analyzed to validate the algorithm. However, due to space limitations, the results of a few cases are presented in the next sub-section.

3.3.1 Inrush Condition

In the simulation, the inrush condition is simulated by energizing the transformer at 0.2 s keeping secondary in open situation. Figure 3.4a shows the effect of inrush on the primary current of a transformer while the secondary current is mostly not in attendance. Figure 3.4b provides the value of arctangent (Δ) of second derivative based differential current and their average value is shown in Fig. 3.4c. It is to be noted from Fig. 3.4c that the average value of angle (θ_{avg}) for inrush condition is considerably high up to 7.8°. This is higher than the set value of threshold in the algorithm, so this situation is identified as an inrush condition.

The tough situation for the protection scheme of the transformer is the case where an internal fault exists in the winding before the transformer is energized. We have performed a test case and observed that the internal fault predominates over the effect of the energization of the transformer.

Thus, the average value of the angle (θ_{avg}) will be always lower than the set threshold. Hence, the proposed algorithm measure this situation as an internal fault condition even though the transformer is switched on in the presence of fault within the transformer. However, in a certain situation (high resistance internal fault), it can be observed that the inrush current waveform is dominant by half a cycle and rapidly

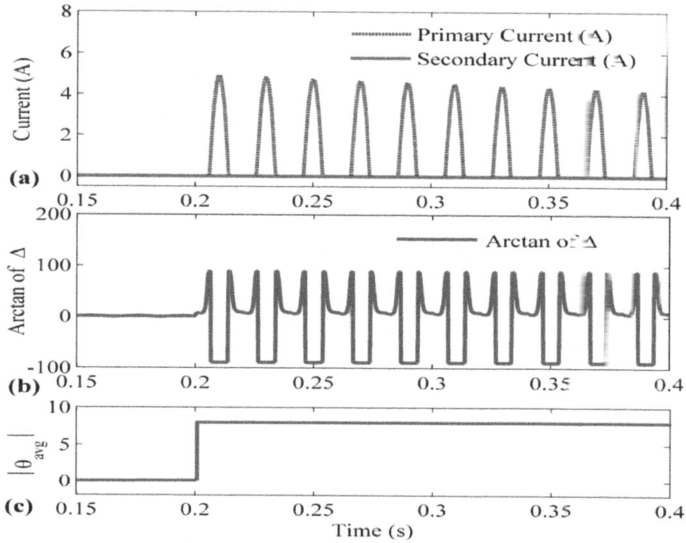

Fig. 3.4 Inrush condition **a** Primary and secondary current of transformer, **b** Arc tan of Δ and **c** average of angle ($\theta_{avg.}$)

changing. This is due to the presence of the fault which leads to an increment in the fundamental component, whereas the DC and second harmonic components are identical to the ones obtained under no-fault conditions at the time of energization. As a result, the magnitude of the average value of angle (θ_{avg}) decays faster in the faulted phase, as compared to a healthy condition. In connection to this situation, the operation of the algorithm is delayed by some time (approximately 50 ms) until the θ_{avg} becomes lower than the set threshold.

Moreover, to check the feasibility of the proposed algorithm to discriminate against the inrush condition and internal fault for unloaded transformer, authors have performed one case in which inrush followed by internal fault. Figure 3.5 shows the result analysis for internal fault simulated during the persistence of inrush conditions. In this case, initially, the transformer is energized at 0.2 s to create inrush and in the presence of inrush situation at 0.3-second internal fault is applied on R-phase of the primary winding of transformer with moderate fault resistance as shown in Fig. 3.5. Figure 3.5a shows a waveform of primary and secondary currents for the simulated case. Figure 3.5b, c exemplify arctangent of Δ and average angle θ_{avg}. It is to be noted that the value of $\theta_{avg} = 7.8$ during a time interval of 0 2–0.3 s and 0.3 s onward it immediately reduces to 0.24°. Thus, the internal fault applied at 0.3 s in the presence of inrush is easily identified.

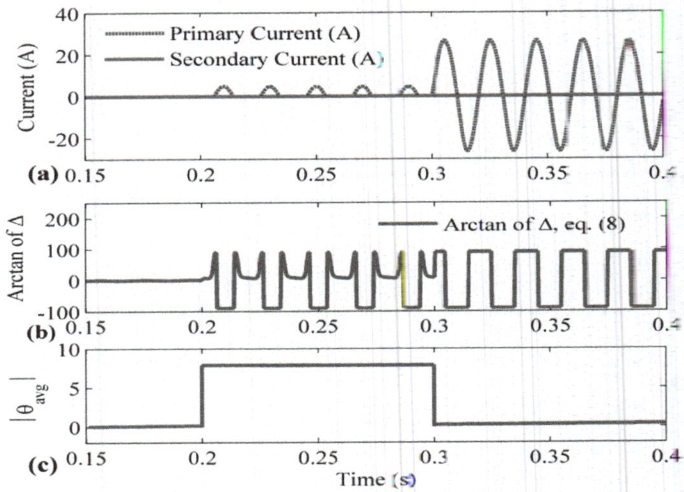

Fig. 3.5 Inrush followed by internal fault **a** Primary and secondary current of transformer, **b** Arc tan of Δ and **c** average Arc tan of Δ ($\theta_{avg.}$)

3.3.2 *Internal Fault in Transformer*

Various internal faults have been simulated on the primary and secondary sides of the transformer at a different percentage of winding including the terminal. Figure 3.6a shows the magnitude of the transformer's primary and secondary current (CT secondary side). Figure 3.6b shows differential current and restraining current magnitude. The differential current becomes greater than restraining current after the fault applied (0.2 s). Also, a biased current 4.028 is higher than the set threshold for the detection of an internal fault. Figure 3.6c illustrates the phasor angle of primary and secondary current and their angle difference. It is observed that the angle difference (θ_d) is 13.04° this indicates an internal fault as the phasors of primary and secondary currents fall within quadrant (<90°).

3.3.3 *High Resistance Internal Fault*

It is worth to check the feasibility of the proposed algorithm for high resistance internal fault condition. To conduct high resistance internal fault on YY connected transformer, deliberately 10 Ω resistances are inserted in the fault path. The fault is being subjected to 30% of the transformer primary winding from the terminal side.

Figure 3.7a demonstrates the difference between primary and secondary current magnitude which is lower than solidly grounded fault (Table 3.2). It is observed that the biased current is higher than the set value in the algorithm. Moreover, the phasor

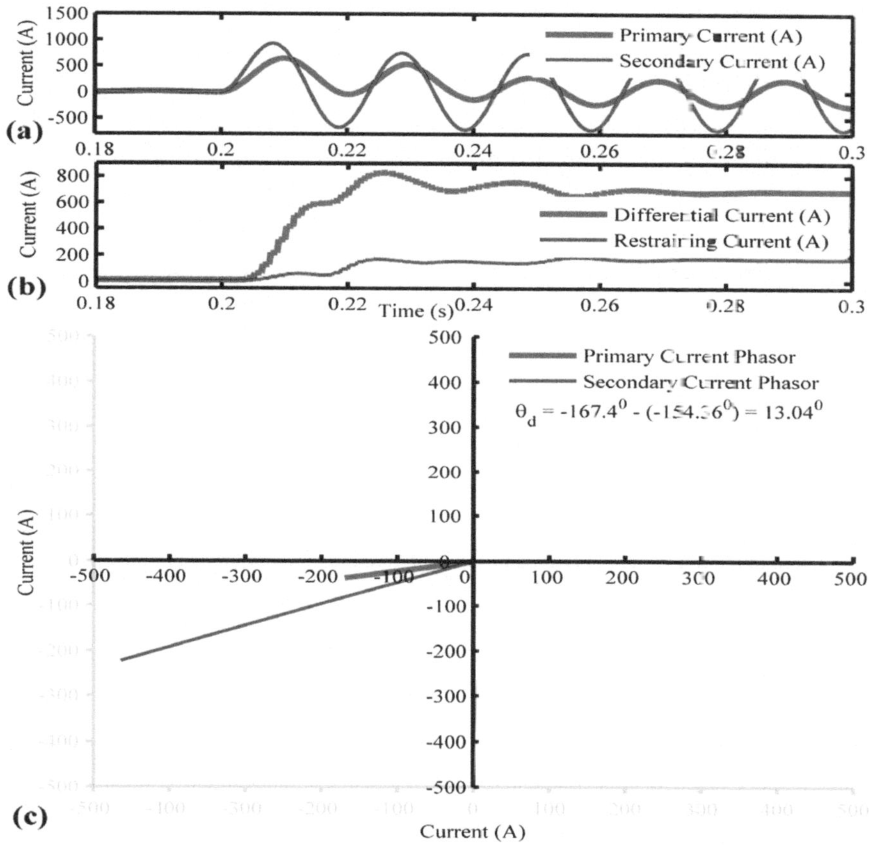

Fig. 3.6 Internal fault **a** Primary and secondary current, **b** differential and restraining current, **c** phasor angle of currents

difference of primary and secondary current falls within the given threshold limit but its margin is moderately increased. Figure 3.7c demonstrates the phasor difference of current during high resistance internal fault. Table 3.2 shows the measured value of phase angles for the said high resistance internal fault condition.

3.3.4 Internal Fault with Heavy CT Saturation

The transformer protection is affected during the CT saturation phenomenon which occurs mainly during a heavy fault condition. An internal fault is simulated on the primary winding with a higher burden on CT secondary to saturate it as shown in Fig. 3.8a. Even though the secondary current of transformer primary CT reduced, the

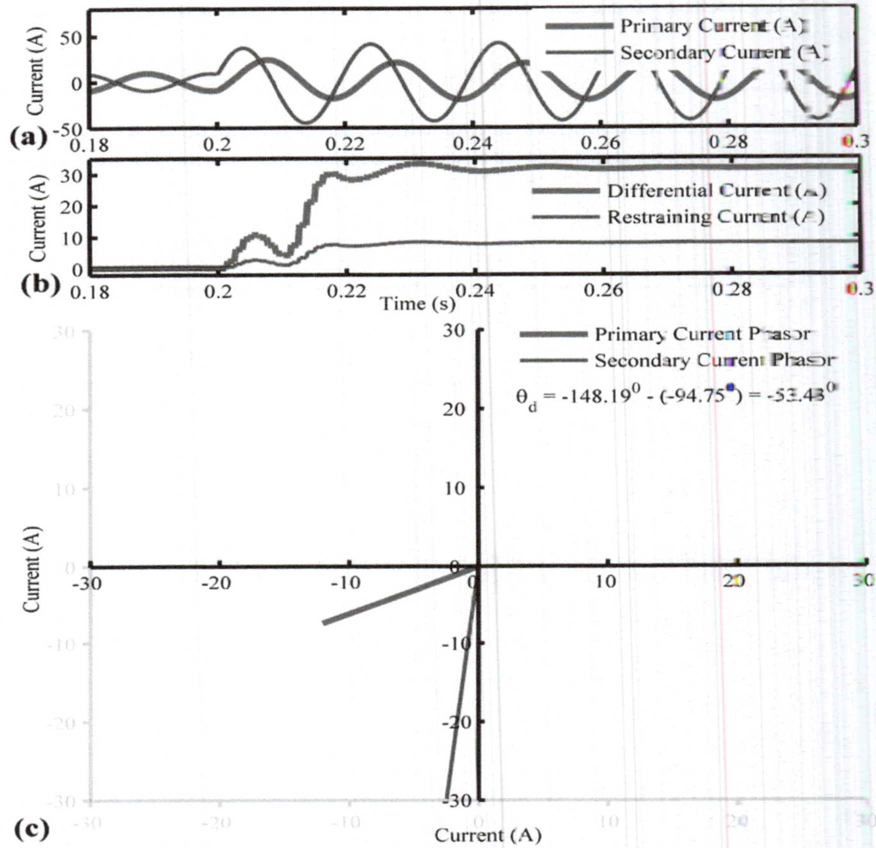

Fig. 3.7 High resistances internal fault **a** Primary and secondary current, **b** differential and restraining current, **c** phasor angle of currents

differential current is greater than the biased current as shown in Fig. 3.8b. Moreover, in this case, the phase angle difference between primary and secondary is slightly increased but remain within the criteria of internal fault detection Fig. 3.8c.

Table 3.2 demonstrates the values of fault current, differential, and restraining current and phasor angle values estimated using the full cycle DFT algorithm. It has been observed from Table 3.2 that the biased current value is much higher than the set threshold as well as the phase angle difference of primary and secondary current is lower than 90°. The angle difference during CT saturation and high resistance increases a little but remains within the limit of the set threshold. As a result the proposed algorithm successfully detects all internal faults including high resistance and issues trip signal.

Table 3.2 Current and phasor comparison of primary and secondary current in internal fault

Sr.	Event	Current comparison					Phasor comparison		
		I_1	I_2	I_d	I_r	I_{biased}	θ_1	θ_2	θ_d
1	Internal fault	173.3	515	688.30	170.85	4.028	−167.4	−154.36	13.04
2	High resistance internal fault	14.14	29.96	44.10	7.91	5.57	−148.19	−94.75	53.437
3	Mild CT saturation	149.94	515	664.94	182.53	3.6429	−150.24	−154.36	4.121
4	Medium CT saturation	111.71	515	626.71	201.64	3.108	−136.94	−154.36	17.416
5	Heavy CT saturation	73.078	515	588.08	220.96	2.6615	−124.91	−154.36	29.453

Where I_1 = Primary Current, I_2 = Secondary Current, I_d = Differential Current, I_r = Restraining Current, I_b = Biased Current, θ_1 = Primary Current Phase Angle, θ_2 = Secondary Current Phase Angle, θ_d = Phasor Difference Between Primary and Secondary Currents

3.3.5 External Fault

Various external faults are created on a 400 kV transmission line and 220 kV bus outside the transformer protection zone. As shown in Fig. 3.9a, during an external fault online at 50 km from the transformer bus, the magnitudes of primary and secondary currents are almost the same with the 180° phase shifted. Also, unlike the internal fault, the restraining current is more than the differential current as shown in Fig. 3.9b. The phase angle difference between primary to secondary is shown in Fig. 3.9c which are displaced approximately 180°. It is to be noted that the algorithm is validated as per set criteria and the trip signal is not issued. Thus, the proposed scheme remains stable during external fault conditions.

3.3.6 External Fault with Heavy CT Saturation

The protection scheme must remain stable during external faults as well as during the CT saturation phenomenon. The CT saturation is obtained by CT model block available in PSCAD/EMTDC. By changing the CT secondary burden resistance, different degrees of CT saturation can be obtained. The performance of the proposed scheme during CT saturation is carried out by simulating different close-in external faults on the primary side of star-star transformer just behind CT location. A line to ground fault is simulated at 200 m from 400 kV transformer-bus with CT secondary

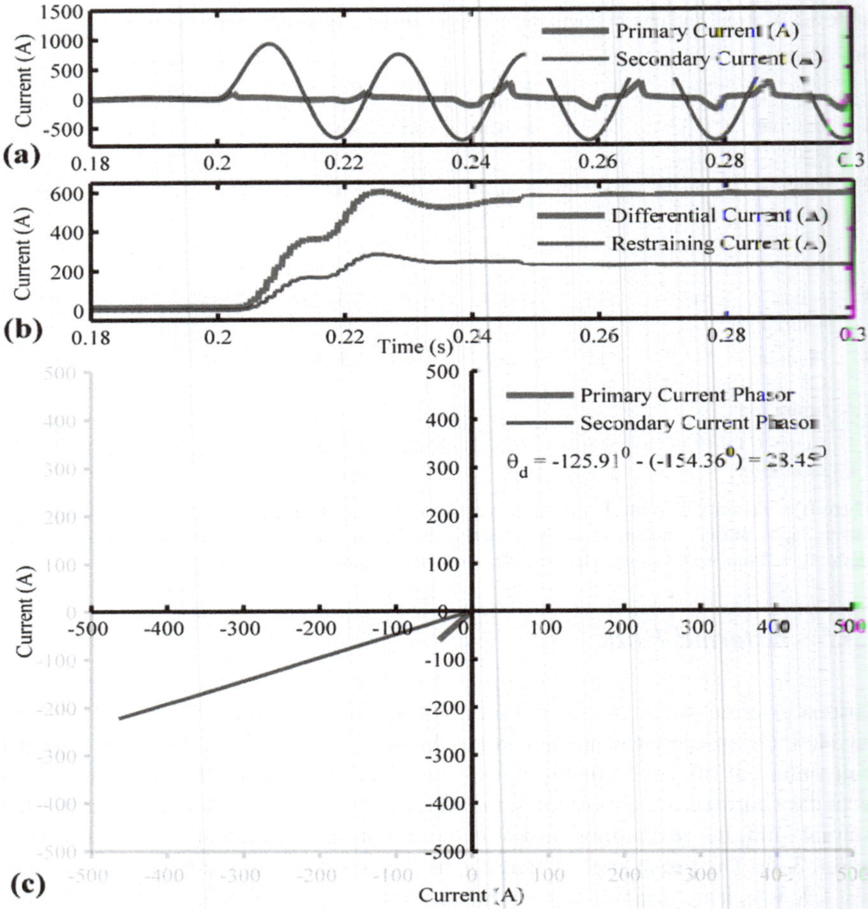

Fig. 3.8 Heavy CT saturation in internal fault **a** Primary and secondary current, **b** differential and restraining current, **c** phasor angle of currents

burden resistance of 15 Ω. Figure 3.10 shows the CT secondary current during the said saturation condition.

As shown in Fig. 3.10a, deep saturation of CT starts just after the point of fault inception (0.2 s), and it results in a substantially higher value of I_d. It is observed that as the severity of saturation increase from low to medium to heavy, the differential current will also gradually increase. Hence, during heavy CT saturation conditions, the only magnitude-based biased differential protection scheme may mal-operate and leads to unnecessary tripping of the transformer in external fault. On the other hand, the phase angle comparison based scheme works well and maintains the phase angle difference above the set value of threshold (>90°) as shown in Fig. 3.10c. Thus,

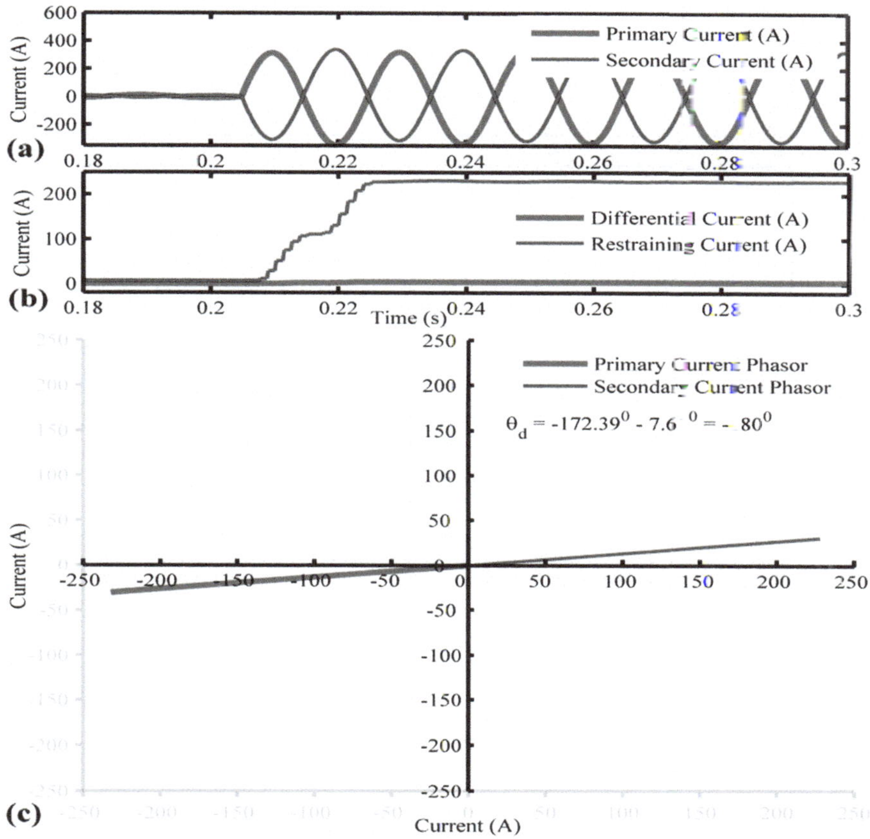

Fig. 3.9 External fault **a** Primary and secondary current, **b** differential and restraining current, **c** phasor angle of currents

combined biased differential and phase angle comparison based scheme successfully detect all kind of external fault and remains inoperative.

Table 3.3 shows the magnitude and phase angle values of current for various external fault conditions. It has been observed from Table 3.3 that calculated biased differential current always remains well below the set threshold for all external faults excluding heavy CT saturation. Conversely, the phase angle differences between primary and secondary current never fall within 90° for all external fault cases.

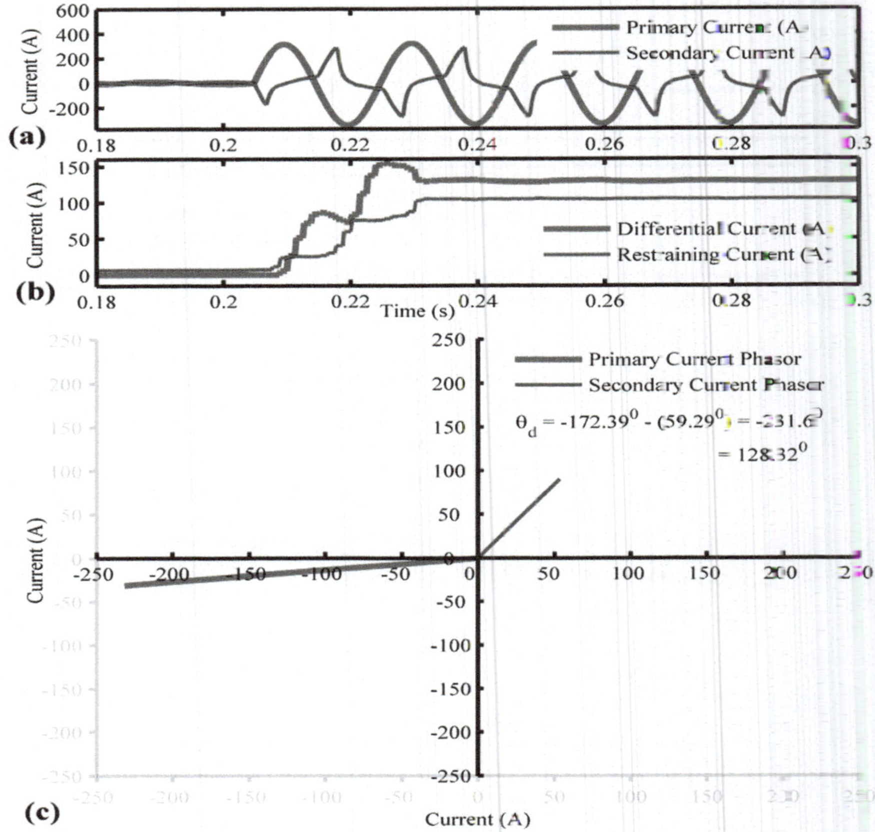

Fig. 3.10 Heavy CT saturation in external fault **a** Primary and secondary current, **b** differential and restraining current, **c** phasor angle of currents

3.4 Experimental Test Setup

3.4.1 Laboratory Prototype

To validate the proposed transformer protection scheme a prototype is developed in a laboratory environment and numerous test cases are conducted. Due to practical limitations instead of using a three-phase transformer, 2 kVA, 230/110 V single phase transformer with multiple tapping is considered in the experiment as shown in Fig. 3.11. Various types of internal on tapping of transformer and external fault are generated with 12 A, 18 Ω variable rheostat.

To develop the hardware setup, single-phase 230 V, 50 Hz local electricity supply is used as the main source. This is given by 0–300 volts AC variac to the test circuit. As a load, 350 Ω, a 2.2 A variable rheostat is connected and an additional two

Table 3.3 Test current and phasor comparison of primary and secondary current in external fault

Sr.	Event	Current comparison					Phasor comparison		
		I_1	I_2	I_d	I_r	I_b	θ_1	θ_2	θ_d
1	External fault	−230.16	234.15	3.99	232.16	0.0172	−172.39	7.608	180
2	High resistance internal fault	−9.908	10.083	0.175	9.995	0.01751	171.73	−8.27	180
3	Mild CT saturation	−228.42	234.16	5.74	231.29	0.02481	−169.23	10.778	180
4	Medium CT saturation	−163.79	234.16	70.37	198.97	0.35364	−169.23	19.494	188.7
5	Heavy CT saturation	−104.43	234.16	130.12	169.09	0.76953	−169.23	45.350	214.5
6	Overloading condition	−43.5	42.76	0.74	43.13	0.0171	120.34	−59.66	180

Where I_1 = Primary Current, I_2 = Secondary Current, I_d = Differential Current, I_r = Restraining Current, I_b = Biased Current, θ_1 = Primary Current Phase Angle, θ_2 = Secondary Current Phase Angle, θ_d = Phasor Difference Between Primary and Secondary Currents

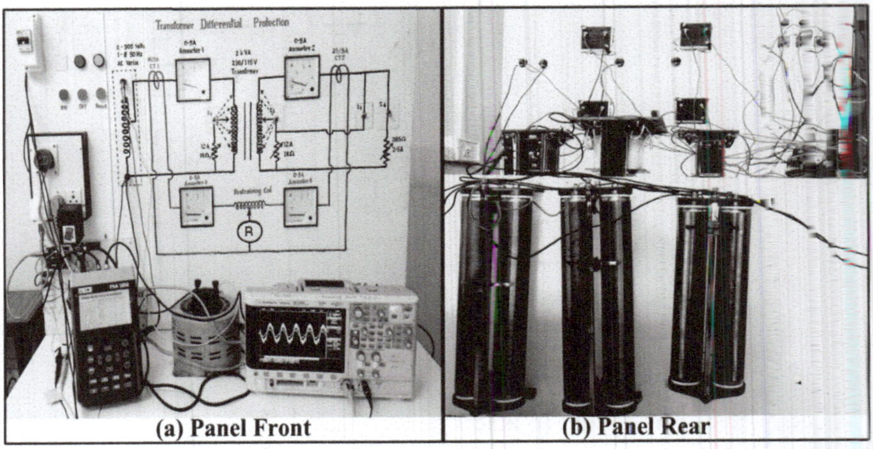

Fig. 3.11 Prototype model developed in laboratory for transformer protection

number of 12 A 18 Ω variable rheostats are connected to generate internal as well as an external fault. Contactors are taken as a circuit breaker (CBs) and current sensor ACS712ELCTR-30A-T is used in hardware to scale down and sense the current in the secondary path of CT. Dedicated Digital Signal Controller (DSC), AVR ATmega 328P as computational hardware is employed in the present work for implementation of the protection scheme. ATmega 328P is also equipped with a large memory capacity of 2 K words of on-chip SARAM, 32 K words on-chip flash memory, and 64 K words off-chip SARAM memory that is sufficient to store large program [25]. The high-performance, 10-bit, 8 channels analog-to-digital converter (ADC) has a minimum conversion time of 500 ns. For the execution of an algorithm, code written in 'C' language using an embedded coder toolbox available in MATLAB is loaded in the memory of the processor. The communication between PC and DSC is done by programmable Universal Asynchronous Receiver Transmitter (UART) which is used to monitor the real-time measurements. The current sensor transfers the current signal into equivalent 5-volt signals. Both the primary and secondary current sensor sends a signal to ATmega328 Microcontroller. In the proposed hardware, primary and secondary internal faults are generated through S_1 and S_2 switch respectively. External faults are created through switch S_3 and load is connected through switch S_4.

Load and fault resistance are variable so modifiable fault current will be made as per the requirement. Additional 18 Ω, 12 A rheostat is connected on the secondary of CT to commence saturation effect during internal as well as an external fault. An algorithm based on a phasor angle and biased differential current is executed in the ATmega328 micro-controller. It is to be noted that the proposed algorithm is validated for all internal fault cases and issues trip signal on the output port within 20–24 ms. To record and compare the waveform of CT secondary current sensor ACS712ELCTR-30A-T, a high resolution four-channel digital storage oscilloscope

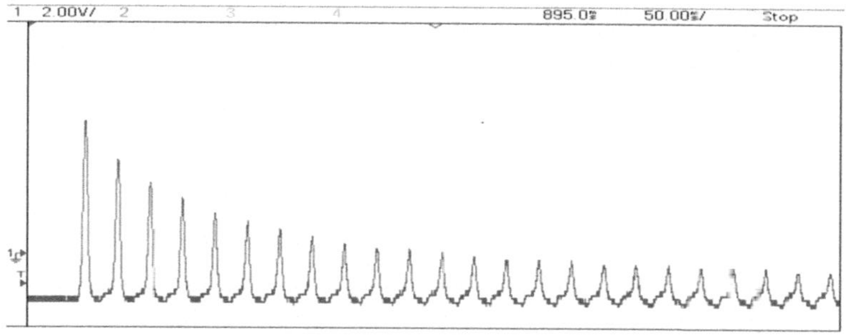

Fig. 3.12 Magnetising inrush

(DSO) is used. To examine the graphical representation of magnitude and phasor difference between primary and secondary current, power and harmonic analyzer PHA5850 is put in operation. Parameters related to hardware set up are illustrated in Appendix.

3.5 Prototype Result Analysis

3.5.1 Inrush

It is observed that when the primary side is connected with supply with the open secondary of the transformer, at around instant of primary voltage zero crossings and with the same polarity of remanence of the core, very severe inrush generated in the primary winding. During the inrush condition, magnetism is a nearby "knee" region of the hysteresis characteristic loop. Due to magnetizing properties of transformer core, inrush current observed in the primary side. Figure 3.12 shows Digital Storage Oscilloscope (DSO) results based on real-time implementation and data recorded. Once, the inrush condition is detected by the proposed algorithm, it returns to the initial stage for the next data sample collection. Hence, no need to calculate differential/restrain current and phase angle difference.

3.5.2 Internal Fault

Figure 3.13a shows the real-time data recorded using a power analyzer during internal fault applied at 30% of the secondary winding of the transformer. It is observed that I_1 (primary side) having fault current whereas I_2 (secondary side) current is zero due to only load connection. Moreover, the biased differential current and phase angle difference satisfies the defined threshold limit of the algorithm. Figure 3.13b shows a vector

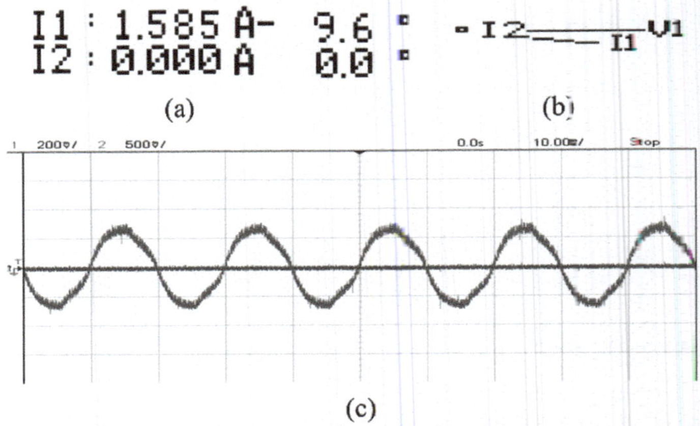

Fig. 3.13 Internal fault **a** Value of primary and secondary current magnitude and phase angle, **b** phasors of primary and secondary current, **c** waveform of primary and secondary current during fault

representation of primary and secondary current phasors. The current waveforms of the primary and secondary side measured in DSO are shown in Fig. 3.13c.

Simulation results (as that of hardware set up): For the validation we have created PSCAD simulation as that of hardware set up in laboratory.

Figure 3.14 shows the results of the proposed algorithm for simulation cases performed in PSCAD and validated in MATLAB. Figure 3.14a shows post fault primary side current (I_1) and secondary current (I_2) during only load connection. Figure 3.14b shows the phasor angle between primary and secondary current which is nearly 7.411°. Thus, the algorithm perfectly detects this situation as an internal fault and its result match with hardware experiments.

3.5.3 External Fault

Figure 3.15a, b, c show the current magnitude, phasors, and waveform of primary and secondary current during external fault created in a laboratory. It is to be noted that the current magnitude is almost the same and the percentage biased current is 0.0724. The phase angle difference of both currents is almost out of phase (180.9°). It is validated that the fault is outside the zone of transformer protection and hence, no trip signal is generated by the algorithm.

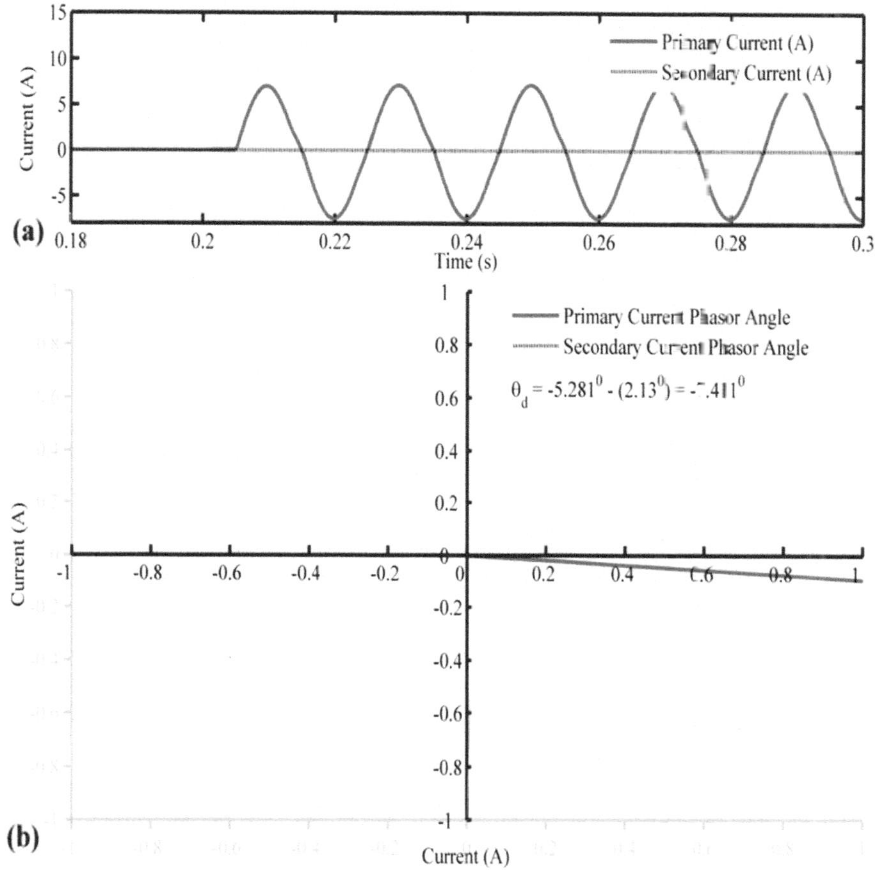

Fig. 3.14 Internal fault **a** Waveform of primary and secondary current, **b** phasor of primary and secondary current during internal fault

3.5.4 External Fault with Deep CT Saturation

To evaluate the performance of the proposed algorithm during heavy CT saturation condition, protective class (5P10) CT having ratio = 10/5 A, burden = 15 VA is used. Further, 18 Ω, 12 A variable rheostat is used as a burden resistance on the secondary side of CT. Figure 3.16 shows the outcome of the proposed test setup during an external fault with the saturation of CT connected on the secondary side of the transformer.

The magnitude-based scheme fails to detect external fault under this situation as the biased current (0.839) goes beyond the set value of the differential principle. Conversely, it has been observed from Fig. 3.16b that the proposed scheme correctly identifies the external fault as the phase angle difference is more than 90° (125.7°).

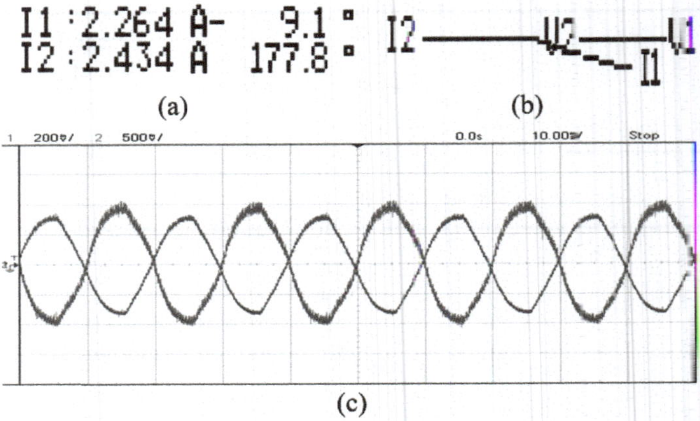

Fig. 3.15 External fault **a** Value of primary and secondary current magnitude and phase angle, **b** phasors of primary and secondary current, **c** waveform of primary and secondary current during fault

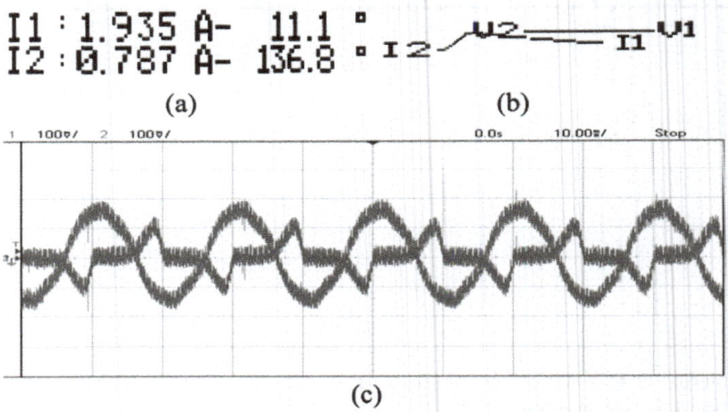

Fig. 3.16 External fault with CT saturation **a** Value of primary and secondary current magnitude and phase angle, **b** phasors of primary and secondary current, **c** waveform of primary and secondary current during CT saturation

Figure 3.16c shows the CT secondary signal in which the transformer secondary current is saturated concerning primary current during an external fault.

Simulation results (as that of hardware set up):

Authors have carried out various external fault on PSCAD simulation and observed that the proposed algorithm successfully identify the situation. Moreover, test results involve CT saturation conditions during an external fault are also performed for validation. Here, the external fault is applied at 0.2 s with higher burden of the secondary side of CT to incorporate saturation effect. CT connected on the secondary side of the transformer gets saturated under external fault due to higher burden.

Figure 3.17a shows a waveform of primary and secondary current. At the same time with the help of the MFCDFT algorithm phasor angle is measured between primary and secondary current which is shown in Fig. 3.17b. Phasor angle difference obtained between primary and secondary side is −142.910, this indicates an external fault condition. Furthermore, the simulation result obtains perfectly matched with the hardware-based experiment.

Various types of fault and abnormal conditions are created and validated on the developed hardware test setup. Table 3.4 shows the phasor angle, percentage biased current measured, and time of operation for all various fault events generated on the prototype.

It is to be noted from the above results and validation that the proposed scheme provides high accuracy during internal fault including high resistance as its performance has been validated both by simulation software and by laboratory prototype.

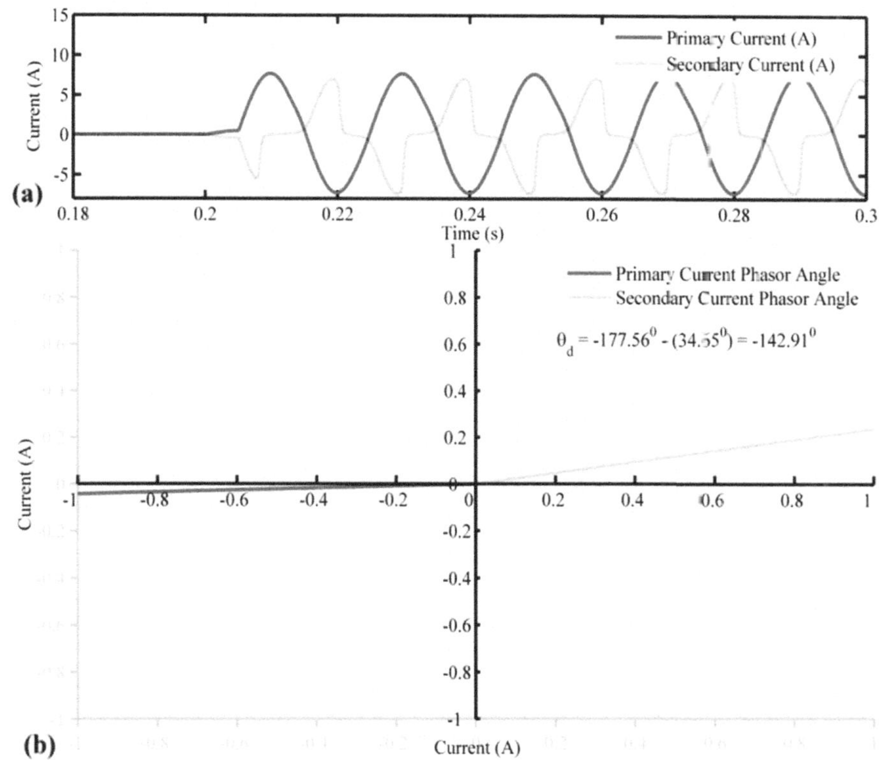

Fig. 3.17 CT saturation under external fault **a** Waveform of primary and secondary current, **b** phasor of primary and secondary current during fault

Table 3.4 Test conditions validation through prototype model

Sr.	Events	Current comparison					Phasor comparison			Time of operation (ms)
		I_1	I_2	I_d	I_r	I_b	θ_1	θ_2	θ_d	
1	Internal fault	1.585	0.0	1.585	0.7925	2	9.6	0.0	9.6	23
2	High resistance internal fault	0.686	0.0	0.686	0.343	2	−30.9	0.0	30.9	24
3	Heavy CT saturation under internal fault	0.415	0.0	0.415	0.2075	2	32.5	0.0	32.5	23
4	External fault	2.26	−2.43	0.17	2.345	0.072	−9.1	177.8	186.9	–
5	Heavy CT saturation under external fault	1.935	−0.787	1.148	1.361	0.839	−11.1	−136.8	125.7	–
6	Overloading condition	0.993	−0.959	0.0334	0.9766	0.034	−16	176	192	–

Where I_1 = Primary Current, I_2 = Secondary Current, I_d = Differential Current, I_r = Restraining Current, I_b = Bias Current, θ_1 = Primary Current Phase Angle, θ_2 = Secondary Current Phase Angle, θ_d = Phasor Difference Between Primary and Secondary Current

The total response time includes data sampling time, the computational time of DSP, and the time required to issue a trip signal from DSP. The logic to decide the response time of DSP is mentioned below.

Response Time:

- The sampling time is decided by considering (i) sampling frequency that is 4 kHz i.e. 0.25 ms (1/4 kHz) and (ii) the number of samples per cycle (80 samples per cycle). Hence, the sampling time of ADC is 0.25 ms × 80 = 20 ms.
- The clock frequency of DSP is 16 MHz. The DSP has pipelined architecture & fast execution time. So, it takes around 0.0625 μs (1/16 MHz) for the execution of an instruction. The size of instruction (proposed algorithm) is a maximum of around 30 kb. Hence, DSP will take around 0.0625 μs*30 kb = 1.875 ms for the execution of the program and generation of the trip signal at its port.
- The propagation delay time of the remaining signal conditioning circuit to be 2 ms.

Hence, the total response time = 20 ms + 1.875 ms + 2 ms, which is approximately in between 23 to 24 ms.

Besides, the proposed scheme provides satisfactory results during severe CT saturation conditions. The same has been validated by implementing the scheme in a laboratory environment and hence, it can be practically put into service for any power system topology.

3.6 Novelty Projected in This Research Work

The proposed scheme utilizes phasor computation techniques and based on phasor values estimated, the algorithm takes the decision of internal fault or external fault condition in power transformer. Looking at this, the proposed technique provides digital operation in a single module which covers protection of power transformer winding, different abnormal conditions. **Hence, the proposed relaying scheme has features like cost efficiency and functional flexibility**.

Moreover, the methods suggested in some of the papers published in the past are tested by generating data in simulating tools/software only. Whereas, in this research work, the proposed technique successfully validated by utilizing real-time practical data from the working model developed in a laboratory setup. **This assures the authentication of the proposed scheme for further implementation in the real field**.

Application of Modified Full Cycle DFT in the proposed phasor angle based fault identification scheme. Furthermore, looking to the originality of the proposed MDFT technique for transformer protection, it has some advantages over existing FFT and DFT based scheme (currently used by many relay manufacturers). This gives reliable and fast operation.

These are:

(1) When a fault occurs, it is desired that the relay used for protection has to respond quickly. The fundamental frequency phasor estimation of the conventional FFT/DFT algorithm is not convergent within a required time limit. Because decaying DC and higher-order harmonics severely inhibit the search for an accurate fundamental frequency signal and delay the convergence time. Whereas, the newly proposed Full Cycle Modified Discrete Fourier Transform (FCMDFT) algorithm has the capability of extracting fundamental frequency components, by eliminating harmonics and the decaying DC offset components, during faults in a system.
(2) This research describes a modified sliding DFT algorithm performs on a sample-by-sample basis whose output rate is equal to the input data rate, with the advantage that it requires fewer computations than the fixed window-based algorithm for real-time protection of the transformer. Hence, in field application the sliding MDFT may be computationally simpler (reduces computational workload) than the traditional FFT/DFT or other filter-based technique [26]. Thus, the proposed scheme has the ability to detect a fault and abnormal conditions in the transformer within required time (speed and accuracy of detection).

Different types of faults and abnormalities are simulated on a transformer and typical simulation results are presented in the manuscript. The simulation results have proved that the proposed technique can perform well whose tripping criterion is based on the exact fundamental frequency component of faulted current waveforms only. Moreover, discrimination of internal fault and external fault decision based on phase angle comparison compare to a lonely differential principle, validation of the proposed scheme on hardware setup and trip decision making are additional strengths of the proposed work. **Thus, the proposed digital transformer protection relay operates efficiently and reliably**.

3.7 Summary

This piece of writing presents a new scheme for the transformer protection based on the average angle of 2nd order derivative of differential current for inrush detection and further discrimination of fault is carried out based on percentage biased differential combined with phase angle comparison between primary and secondary current. Numerous test cases including inrush, overload, fault, and abnormal conditions are generated in PSCAD^{TM} software. The algorithm is developed using a modified full-cycle DFT filter to estimate the magnitude and phase angle of current signals. The decision of trip signal during internal and external fault is taken using AND logic of biased differential and phase angle comparison based technique. Various test conditions such as transformer inrush, overloading, internal faults on the percentage of the transformer winding, high resistance fault, external fault with CT saturation are simulated and successfully validated. Moreover, the algorithm is authenticated on hardware setup developed in a laboratory environment. Simulation and prototype results demonstrate that the proposed algorithm can discriminate an internal fault and other abnormalities perfectly. One of the advantages of this scheme is minimum statistical computation, which is easily applicable in relaying the program and gives the trip signal within 24 ms during an internal fault. Conversely, the proposed scheme remains inoperative during all external faults considering heavy CT saturation.

3.8 Published Article Based on This Work

- Dharmesh Patel, N. G. Chothani, K. D. Mistry, "Discrimination of Inrush, Internal and External Fault in Power Transformer using Phasor Angle Comparison and Biased Differential Principle", Electrical Power Component and System, Tailor and Francis Group, 46(7), pp. 788–801, 2018.

Appendix

Simulation model data:

Source data	3-phase, 300MVA, 400kV, 50Hz
Line data	Length = 80 km, System voltage = 400 kV Positive-sequence impedance = 0.0297 + j0.332 Ω/km Zero-sequence impedance = 0.162 + j1.24 Ω/km Positive-sequence capacitance = 12.99 nF/km Zero-sequence capacitance = 8.5 nF/km
Transformer data	YY connected, 315 MVA, 400/220 kV, 3-phase, with 0.1 pu leakage reactance (220 kV side of the transformer is connected to the infinite bus)
CT data	Primary-1000/5 A, Secondary-1800/5 A, Secondary winding resistance, and inductance = 0.5 Ω and $0.8e^{-3}$ H

Equipment data for hardware:

Transformer data	2 KVA, 230/115 V, 1-phase, 50 Hz, %Z = 12
CT data	Primary side: 10/5 A, 15 VA, 5p10 and for secondary side 20/5 A, 15 VA, 5p10
Load	Lamp load, 25 A
Source data	1-phase, 0–300 V, 50 Hz, Variable supply from the electricity board

References

1. Kulkarni SV, Khaparde SA (2004) Transformer engineering: design and practice. Taylor & Francis, London
2. Hong-ming S, Tao Z, Shao-feng H, Ou L (2014) Study on a Mal-operation case of differential protection due to the interaction between magnetizing inrush and sympathetic inrush. In: 2014 IEEE PES general meeting | Conference exposition, pp 1–5
3. Hunt R, Schaefer J, Bentert B (2008) Practical experience in setting transformer differential inrush restraint. In: 2008 61st Annual conference for protective relay engineers, pp 118–141
4. Hong C, Haifeng L, Hua L, Jiran Z, Haiguo T, Zhidan Z (2017) Waveform complexity analysis of differential current signal to detect magnetizing inrush in power transformer. In: 2017 9th International conference on measuring technology and mechatronics automation (ICMTMA), pp 120–123
5. Barbosa D, Netto UC, Coury DV, Oleskovicz M (2011) Power transformer differential protection based on Clarke's transform and fuzzy systems. IEEE Trans Power Deliv 26(2):1212–1220
6. Tripathy M, Maheshwari RP, Verma HK (2010) Power transformer differential protection based on optimal probabilistic neural network. IEEE Trans Power Deliv 25(1):102–112
7. Mohamed EA, Abdelaziz AY, Mostafa AS (2005) A neural network-based scheme for fault diagnosis of power transformers. Electr Power Syst Res 75(1):29–39
8. de Faria H, Costa JGS, Olivas JLM (2015) A review of monitoring methods for predictive maintenance of electric power transformers based on dissolved gas analysis. Renew Sustain Energy Rev 46:201–209
9. Díaz G, Blanco ID, Arboleya P, Gómez-Aleixandre J (2006) Zero-sequence-based relaying technique for protecting power transformers and its performance assessment using unsupervised learning ANN. Eur Trans Electr Power 16(2):147–160
10. Moravej Z, Abdoos AA, Sanaye-Pasand M (2011) Power transformer protection using improved S-transform. Electr Power Compon Syst 39(11):1151–1174
11. Stanbury M, Djekic Z (2015) The impact of current-transformer saturation on transformer differential protection. IEEE Trans Power Deliv 30(3):1278–1287
12. Ahmadi M, Samet H, Ghanbari T (2015) Discrimination of internal fault from magnetising inrush current in power transformers based on sine-wave least-squares curve fitting method. IET Sci Meas Technol 9(1):73–84
13. Khan UN, Sidhu TS (2014) A phase-shifting transformer protection technique based on directional comparison approach. IEEE Trans Power Deliv 29(5):2315–2323
14. Valsan SP, Swarup KS (2008) Wavelet based transformer protection using high frequency power directional signals. Electr Power Syst Res 78(4):547–558
15. Jettanasen C, Ngaopitakkul A (2012) The spectrum comparison technique of DWT for discriminating between external fault and internal fault in power transformer. J Int Counc Electr Eng 2(3):302–308
16. Naumov VA, Shevtsov VM (2003) Mathematical models of current transformers in algorithms of differential protection. Power Technol Eng 37(2):123–128
17. Eissa MM, Shehab-Eldin EH, Masoud ME, Abd-Elatif AS (2012) Laboratory investigation for power transformer protection technique based on positive sequence admittance approach. Eur Trans Electr Power 22(2):253–270
18. Rahmati A, Sanaye-Pasand M (2015) Protection of power transformer using multi criteria decision-making. Int J Electr Power Energy Syst 68:294–303
19. Noshad B, Ghanavati B, Ahmadzadeh M, Mohammadzadeh S (2015) Control of unusual mal-operation of three-phase power transformer differential protection due to ultra saturation phenomenon based on Clark's transform and discrete wavelet transform. Renew Sustain Energy Rev 51:1276–1287
20. PSCAD Research Center (2005) EMTDC-transient analysis for PSCAD power system simulation. Winnipeg, MB, Canada

21. Chothani NG, Bhalja BR (2014) Development of a New Bus Zone Identification Algorithm Based On Phase Angle Comparison Using Sequence Components Of Currents. Electr Power Compon Syst 42(2):215–226
22. Zadeh MRD, Zhang Z (2013) A new DFT-based current phasor estimation for numerical protective relaying. IEEE Trans Power Deliv 28(4):2172–2179
23. Thompson MJ (2011) Percentage restrained differential, percentage of what? In: 2011 64th Annual conference for protective relay engineers, pp 1–12
24. Kasztenny B, Kulidjian A, Campbell B, Pozzuoli M (2000) Operate and restraint signals of a transformer differential relay. In: 54th Annual Georgia tech protective relaying conference, 3–5 May 2000, pp 1–24
25. Atmel (2016) ATmega328/P. AVR microcontrollers, p 442
26. Jacobsen E, Lyons R (2003) The sliding DFT. IEEE Signal Process Mag 20(2):74–80

Chapter 4
Adaptive Digital Differential Protection of Power Transformer

Due to the presence of Distribution Generation (DG), the power system becomes more complicated and stability of power is the main challenging task Saturation of Current Transformer (CT) imposes a great dilemma on differential relaying scheme. This manuscript presents a new differential algorithm for distribution transformer protection which adaptively set its characteristic in the event of CT saturation. The proposed scheme is capable to detect magnetizing inrush condition, high resistance internal fault and discriminate external fault with CT saturation. The validation of the proposed scheme is done by simulating a part of the power system in PSCAD™ software and programming in MATLAB software. A Full Cycle Discrete Fourier Transform (FCDFT) is implemented to validate the differential protective scheme for 15 MVA, 66/11 kV distribution transformer. An adaptive concept of the differential characteristic is employed in the algorithm to maintain the stability of relay during external fault with CT saturation. Validation and authenticity of the proposed technique are carried out with various test conditions generated under wide variation in system parameters. The result on 2 kVA, 230/110 V, single-phase transformer shows that the proposed scheme is capable to discriminate inrush, internal and external fault also with CT saturation conditions.

4.1 Literature Studied on Transformer Protection

Nowadays reliability and continuity of supply are the main issues in the power system. The distribution transformer is one of the most important equipment to transfer power from one voltage level to another in a power system. Due to different voltage ratio, current ratio and other constraints, protection of distribution transformer face problems like CT saturation and magnetizing inrush. Even when a fault occurs in the power system, exponential D.C. component results to distort secondary current and malfunctions in relaying operation. These effects generate a major issue to discriminate against internal and external faults in transformer protection.

D. Patel and N. Chothani, *Digital Protective Schemes for Power Transformer*, Power Systems, https://doi.org/10.1007/978-981-15-6763-6_4

Various methods are introduced by researchers for the detection of inrush conditions, internal and external fault conditions. Most of them utilize 2nd harmonic component values for magnetizing inrush detection. Switching instant is the main function of the generation of inrush in a transformer. Abdel et al. [1] proposed a strategy to reduce inrush current in power transformer based on switching instant selection in a single-phase PV system. Hooshyar and Sanaye-Pasand [2] measured decaying DC offset and CT saturation in fault current using least error squares technique with higher immunity against noise and harmonics on prototype work. Samet et al. [3] proposed discrimination techniques for internal fault and inrush conditions in transformer protection based on the function of the autocorrelation technique and compare results with various techniques. The frequency response of mechanical deformation is analyzed based on transfer function estimation by Narayana et al. [4] which provide information before and after fault condition and generate the possibility of maloperation during CT saturation and inrush conditions. Many classifiers and decomposing techniques are also utilized for transformer protection. Valsan and Swarup [5] proposed transformer protection using power directional signals based on wavelet transform decomposing technique. Bigdeli et al. [6] elaborated transformer winding fault with the transfer function based SVM classifier method. Galdi et al. [7] explained the Genetic Algorithm (GA) working on the load current and the measured hot-spot temperature pattern in the transformer. Ahmadi et al. [8] proposed Least Square curve fitting based discrimination of internal fault and inrush conditions by considering CT saturation conditions. Wei et al. [9] presented incipient fault detection in oil-immersed power transformer using least square support vector machine (LSSVM) optimized with particle swarm optimization (PSO) on the dissolved gas analysis (DGA).

Ozgonenel [10] estimated CT saturation with X/R ratio and phase angle comparison based methodology which provide a basic idea for detection of CT saturation and apply adaptive criteria on transformer protection to fulfill all aspects. Stanbury and Djekic [11] suggested CT saturation impact on transformer protection with three parallel 132/11 kV transformers. Contradictory, during close in an external fault condition, the system may mal-operate. Hajipour et al. [12] explained CT saturation compensation in transformer differential relays with noise immunity function. But, the response time along with CT saturation detection is not clear. Various methods of CT saturation detection have been suggested which aid in accurate power system protection. Ajaei et al. [13] elaborated CT saturation effect and its compensation accurately with an estimation of current phasor within very short time compensation is applied. Bak et al. [14] demonstrated first-order derivatives of the signals as coordinates of a 3D vector and second-order derivatives are utilized for the detection of CT saturation. Shi et al. [15] presented CT saturation compensation with partial nonlinear model and assume remnant flux is zero at the instant of the fault and validate data at the various testing condition. Kang et al. [16] elaborated on the wavelet transform based CT saturation detection algorithm, so far estimation of saturation index for identification is very complicated even scaling factor also varies with the protective scheme. Esmail et al. [17] suggested partial CT saturation based on first derivative and waveform compensation for the current transformer with a generic discrimination index and reconstruction of the saturated waveform by Kalman filtering. Though, the methods proposed above for CT saturation have not considered various tests like fault resistance, different CT burden, and power factor. Moreover, the least estimation and regression-based technique depended on variables and parameter calculation which reduces sensitivity and observes an error in the LES method. Finally, adaptive protection considering CT saturation detection gives a better option to improve transformer protection with 2nd harmonic component-based magnetizing inrush detection. Kang et al. [18] elaborated compensation of CT saturation for power transformer. However, the time of operation is so high (42 ms) for the detection and compensation of CT saturation.

This perspective present discrimination of internal and external fault conditions based on the biased differential principle. It provides adaptive protection in the event of CT saturation during external fault and unwanted circumstances. MFCDFT algorithm is used to implement biased differential protection and the third-order derivative method is applied to detect saturation of CT. The algorithm is constructed simply and based on the easy implementation in the practical field. The basic operating characteristic of the differential relay is set at 30% and able to shift up to 70%, which covers the maximum limit of CT saturation. The scheme is validated under various test conditions and provides high stability against external abnormalities in the transformer.

4.2 Problem Discussion and Definitions

Percentage differential relay is normally utilized for unit protection of a transformer, generator, and busbar. Biased differential characteristics are divided into two to three slopes as per the demand of accuracy. Figure 4.1 shows a simple two-stage characteristic of the biased differential relay. Portion "OA" is decided as the restraining portion which covers 5% of normal rated secondary current which is considered as

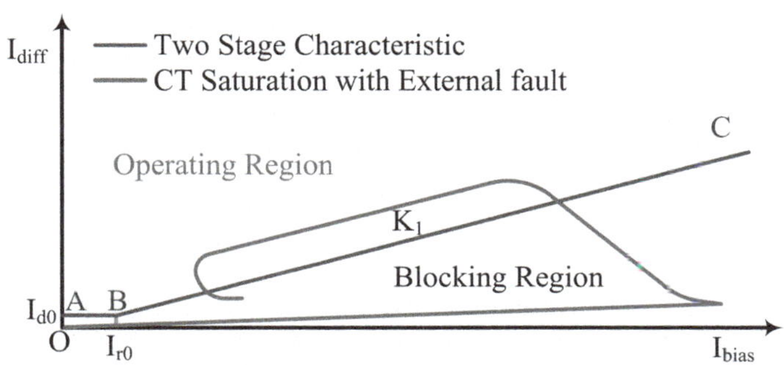

Fig. 4.1 Two-stage biased differential relay characteristics

a basic setting i_1–i_2 covered by “AB”. Slope K_1 is considered as bias setting 120% of rated secondary current with normally 30% slope covered by BC [19].

Generally, the setting of K_1 varies from 0.4 to 0.7 for the restraining current of 0.2–0.6 times I_{rated} and operating current of 0.8–1.0 times I_{rated} considering single slope characteristics. The actual trajectory of I_{diff}/I_{bias} crosses the locus of biased differential characteristic during all internal fault and external fault with CT saturation. Figure 4.1 show the trail of I_{diff}/I_{bias} trajectory crosses the line BC (differential characteristic), during external fault with CT saturation condition. This indicates that the relay may mal-operate if saturation condition is not detected within the time during an external fault.

CT saturation is the main distortive effect of maloperation of protective schemes under normal and abnormal conditions. To implement proper CT saturation detection by providing adaptive criteria in percentage bias differential protection schemes itself is a complicated process. This paper provides a new algorithm to avoid problems due to CT saturation with the adaptive relaying concept. Mild, medium, and heavy saturation effects are considered for validation of an algorithm under internal as well as an external fault condition. The detailed algorithm is explained in Sect. 4.4.

4.3 System Modeling

The proposed single line diagram for testing is as shown in Fig. 4.2, distribution transformer is connected between 66 and 11 kV bus with 15 MVA rating. 66 kV line is connected to 66 kV line is connected to Thevenin’s equivalent multi-machine system through GT. 11 kV distribution feeder is connected to load and distribution generations. For the validation of a transformer protection system, CT is connected to the primary and secondary sides of the transformer. The model is validated by PSCAD™ software [20]. For accurate measurement of magnetic characteristics, JA model type CTs are used, to carry out a specific simulation on transformer protection.

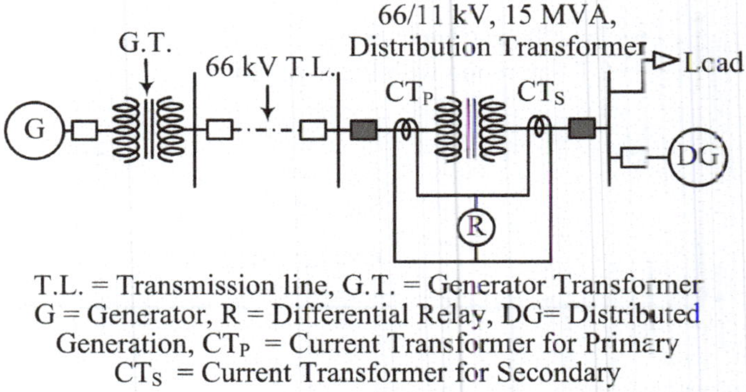

Fig. 4.2 Line diagram for testing

Many test cases are evaluated for the analysis of the proposed algorithm. All test cases like internal and external (L-G, L-L, L-L-G, L-L-L, L-L-L-G) fault are generated through inbuilt fault simulation block on various test conditions like a burden on CT with different remnant flux, fault resistance, fault location. For accurate measurement of sampled current data, JA (Jiles-Atherton) model type CTs are used in the simulation. The system parameter is given in Appendix.

4.4 Proposed Adaptive Relaying Scheme

During magnetizing inrush condition, normally fundamental component, DC component, 2nd harmonic, 3rd harmonic, 4th harmonic, and 5th harmonic components are present as a percentage of 100%, 55%, 63%, 26.8%, 5.1%, and 4.1% sequentially [21]. The magnetizing inrush current is in reach of the second harmonic current component. Thus, based on the 2nd harmonic and fundamental current ratio, the magnetizing condition in the transformer is detected. Modified full-cycle Discrete Fourier Transform (MFCDFT) is used to extract the fundamental and all other harmonic components from no load to full load to faulty current signals. The full cycle MDFT algorithm which can extract exact fundamental frequency components from a given input signal is presented in this study. Consider a full cycle time period T and continuous sinusoidal (current) signal *f(t)* which contains DC component and $N-2$ order harmonics. If the sampling frequency is considered as f_s then N is the sampling rate for a fundamental frequency period. The sample period/time step of the algorithm is $\Delta T = T/N$ (400 μs in this study). Then, the *f(t)* and the Kth sample signal *f(k)* are represented by Eqs. (4.1) and (4.2).

$$f(t) = A_0 + \sum_{n=1}^{N-2} A_n \cos(n\,\omega t + \theta_n) \tag{4.1}$$

$$f(k) = A_0 + \sum_{n=1}^{N-2} A_n \cos\left(\frac{2nk\pi}{N} + \theta_n\right) \tag{4.2}$$

Fundamental frequency complex phasor contains both, real part $F_{r(k)}$ and imaginary part $F_{i(k)}$.

$$F_{r(k)} = \tfrac{2}{N} \sum_{r=k-N+1}^{k} * f(r) * \cos\left(\frac{2r\pi}{N}\right) \tag{4.3}$$

$$F_{i(k)} = \tfrac{-2}{N} \sum_{r=k-N+1}^{k} * f(r) * \sin\left(\frac{2r\pi}{N}\right) \tag{4.4}$$

Equations (4.3) and (4.4) represent FCDFT. Hence, when K ≥ N following equations are obtained:

$$F_{r(k)} = A_1 \cos\theta_1 \tag{4.5}$$

$$F_{i(k)} = A_1 \sin\theta_1 \tag{4.6}$$

$$\text{Amplitude,}\ \ A_1 = \sqrt{F_{r(k)}^2 + F_{i(k)}^2} \tag{4.7}$$

$$\text{Phase angle,}\ \ \theta_1 = \tan^{-1}\left(\mathrm{F_{i(k)}} / \mathrm{F_{r(k)}}\right) \tag{4.8}$$

The calculated 2nd harmonic and fundamental phasor values of given input (current) is further utilized for adaptive percentage biased differential algorithm.

Moreover, the 3rd order derivative of CT secondary current gives useful information regarding CT saturation condition in a power system [22].

4.4.1 Third (3rd) Derivative-Based Technique

The secondary current of CT is given by Eq. 4.9.

$$I_{s(n)} = X.e^{nt/T_s} + Y.e^{nt/T_p} - Z.\sin\left(\frac{2\pi}{N}n - \alpha - \beta\right) \tag{4.9}$$

where, T = Time constant, N = Numbers of samples per cycles, α = Voltage angle at the instant of fault occurrence, β = angle introduced due to CT secondary parameters, X, Y, Z are constant parameters, S and P refers as secondary and primary of CT, n is the recent sample,

The first difference between $I_{s(n)}$ is defined as

$$\delta_{1(n)} = I_{s(n)} - I_{s(n-1)} \tag{4.10}$$

Here exponential terms in $I_{s(n)}$ are reduced and become negligible as the time constant is large.

The second differential equation of $I_{s(n)}$ is defined as

$$\delta_{2(n)} = I_{s(n)} - 2I_{s(n-1)} - I_{s(n-2)} \tag{4.11}$$

A third differential equation is

$$\delta_{3(n)} = I_{s(n)} - 3I_{s(n-1)} + 3I_{s(n-2)} - I_{2(n-3)} \tag{4.12}$$

Saturation detection with 1st to 3rd derivative-based technique is given by index

$$\delta_n = \frac{1}{H}\left[\frac{\delta_{1(n)}}{1} + \frac{\delta_{2(n)}}{2} + \frac{\delta_{3(n)}}{3}\right] \text{ Where H is the sampling interval} \tag{4.13}$$

To detect CT saturation during different fault cases, a certain threshold value is compared with a third derivative-based technique as derived in above Eq. 4.13. Hence, an adaptive threshold is estimated for saturation detection as below,

$$A_s = F * \sqrt{2} * .I_{f(\max)} * \left[2 * \sin\left(\frac{\pi}{N}\right)\right]^3 \tag{4.14}$$

where F = safety factor, N = numbers of samples, I_f = fault current amplitude

Now, by comparing the saturation index derived from Eq. 4.13 with the adaptive threshold i.e. Eq. 4.14, it is observed whether saturation occurs or not [22]. Further, the degree of CT saturation (X_s) can be obtained as below which is useful to shift (modify) the biased differential characteristics as per desire.

Hence, when $\delta_n \geq A_s$ then CT gets saturated.

$$\begin{aligned}\text{Degree of saturation} = \mathrm{X_s} &= \frac{\delta_n - A_s}{\delta_n} * 100\% \\ &= \left(1 - \frac{A_s}{\delta_n}\right) * 100\end{aligned} \tag{4.15}$$

With the use of $\mathrm{X_s}$, the slope of relay characteristics is determined as,

$$K_1 = 0.3 + \Delta M_s \tag{4.16}$$

where $\Delta M_s = 0.9X_s$

In Eq. 4.16, the initial slope setting of 0.3(30%) is considered as a reference slope incorporating the effect of CT error, relay measurement error, and on-load tap compensation error [23]. Here, ΔM_s is the relative slope step and it depends on the degree (level) of CT saturation (X_s). Hence, based on the value of the CT saturation level, the slope characteristic (K_1) will shift adaptively from low percentage K_1^1 (30%) to maximum K_1^n (70%) during external fault with CT saturation condition.

Figure 4.3 shows a detailed flowchart of the proposed adaptive differential protection scheme for transformer protection. This algorithm is divided into four stages. Detection of inrush condition, on the bases of the ratio of 2nd harmonics to the fundamental component, must be greater than 20%. The second stage demonstrates the calculation of the saturation index of primary and secondary current as per Eq. 4.13 and adaptive threshold. The third stage is fault determination and discrimination whether it is external or internal, based on differential and bias current estimated using MFCDFT. The fourth stage is the implementation of the adaptive criteria for CT saturation based on the saturation index as per Eq. 4.15 and 4.16. The effect of CT saturation during external fault will modify the slope of biased differential

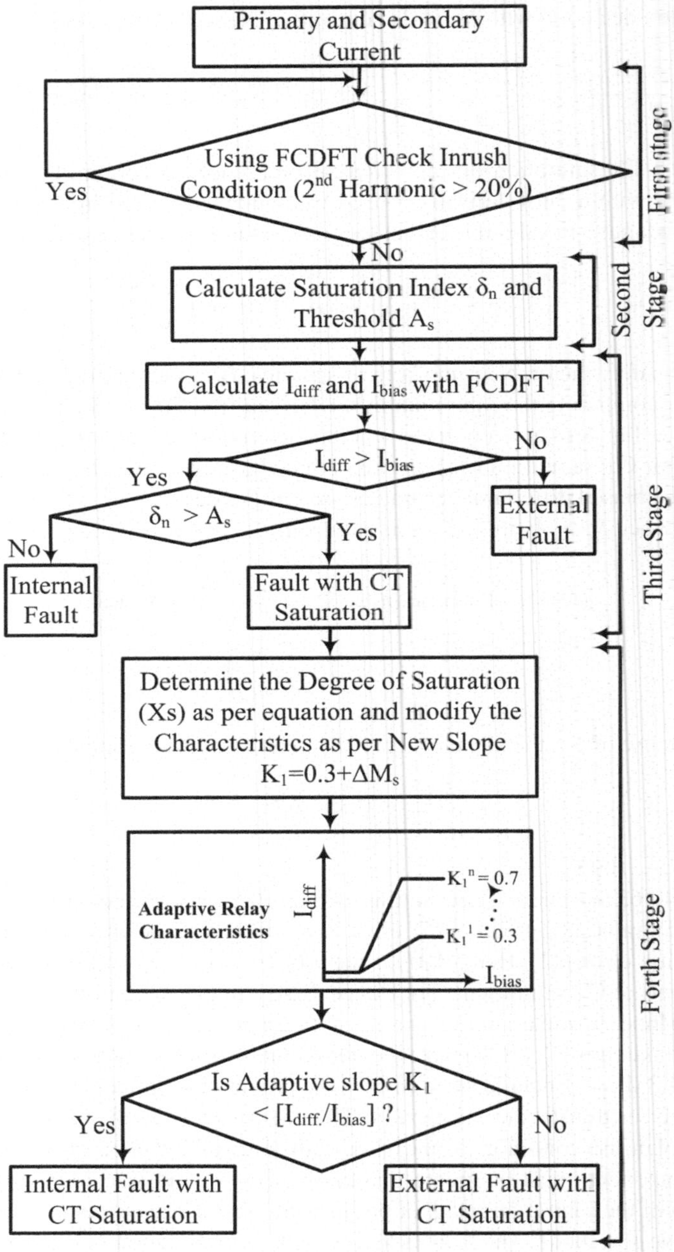

Fig. 4.3 Proposed fault zone identification algorithm

characteristics from 30% to 70%. However, the trip signal will be issued only when the ratio of I_{diff} to I_{bias} settles above the adapted slope K_I (an internal fault with CT saturation) or otherwise it is blocked (an external fault with CT saturation). Thus, the proposed algorithm effectively distinguishes the internal and external fault even under CT saturation conditions.

4.5 Result Discussion

Various test cases are simulated on the considered distribution transformer of Fig. 4.2. All internal and external faults are applied at 0.2 s once the transformer is fully energized. To validate the proposed scheme, different cases are generated considering light saturation to heavy saturation of CT during external faults. Moreover, magnetizing inrush condition, full load condition, high resistance internal fault, and external fault conditions are simulated to check the practicability of the proposed algorithm. The following subsection exemplifies result of various test cases

4.5.1 Magnetizing Inrush Condition

Figure 4.4a shows the effect of magnetizing inrush current when the transformer is energized from the primary side (66 kV) and secondary is kept under no-load condition. The 2nd harmonic component calculated by FCDFT is very large under inrush condition. As per algorithm, if the 2nd harmonic component increases more than 20% of fundamental than this situation is considered as a magnetizing inrush. Figure 4.4b shows the comparison of 2nd harmonic and fundamental components separated during the initial operation of the algorithm. It is to be noted that during the inrush condition the algorithm returns to its data acquisition unit. If the inrush condition is not detected, the algorithm follows the next step as per Fig. 4.3.

4.5.2 Internal Fault on Transformer Winding

During internal fault on a transformer, the primary and secondary current phasor angle is almost in-phase as shown in Fig. 4.5a. It is observed that differential current (I_{diff}) is greater than the restraining current (I_{bias}) as shown in Fig. 4.5b. Moreover, the differential v/s bias current trajectory crosses the set biased differential characteristics as shown in Fig. 4.5c, thus relay successfully issue trip signal. A high resistance internal fault is also simulated with $R_f = 10\,\Omega$, under this condition proposed scheme gives an accurate result with a minimum time of operation. Various system parameters such as types of fault, fault on transformer winding, and fault resistance are considered using internal fault for validation of the proposed scheme.

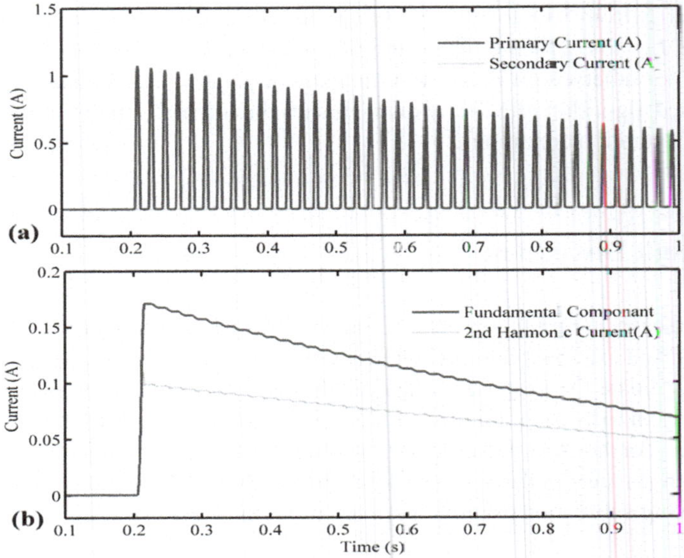

Fig. 4.4 Magnetizing inrush condition, **a** primary current and secondary current, **b** fundamental and second harmonic components

Table 4.1 shows results for various internal faults carried out on the transformer winding in terms of operating time. It is to be noted from Table 4.1 that the total response time of the proposed algorithm is about 31–35 ms. This includes time delay involved in data sampling plus time for all other computations carried out.

4.5.3 Transformer Internal Fault with CT Saturation

During an extreme internal fault condition, the CT may get saturate and results in the lower differential current but still greater than biased current. The detection of CT saturation may mal-function the algorithm as the slope will adaptively be shifted depending on the degree of saturation.

To check the practicability of the proposed scheme, authors have simulated various test cases of internal fault with CT saturation. Figure 4.6 illustrates the validation of the proposed scheme for internal fault with moderate to heavy CT saturation. Though the differential current is greater than the biased current the third stage of the algorithm (Fig. 4.3) detects a fault with CT saturation. Accordingly, the relay will adjust the characteristic based on the level of saturation estimated during a fault. Nevertheless, the differential v/s bias current trajectory crosses the modified slope K_1^n as shown in Fig. 4.6, c, d, thus relay successfully issue trip signal.

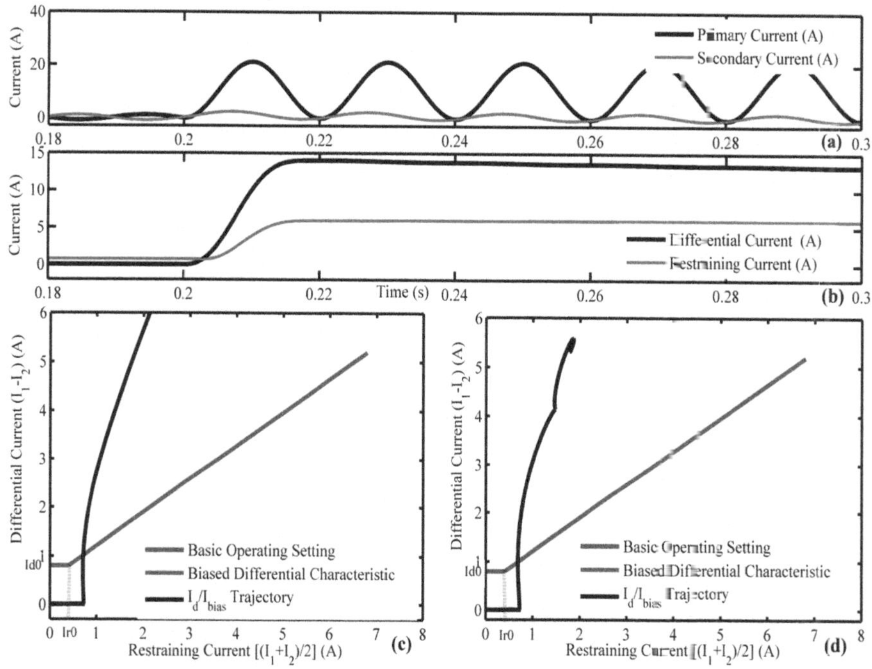

Fig. 4.5 Internal fault, **a** primary versus secondary current, **b** magnitude of differential and restraining current, **c** I_{diff}/I_{bias} trajectory without fault resistance, **d** I_{diff}/I_{bias} trajectory with 10 Ω fault resistance

4.5.4 External Fault Condition

Various external faults are simulated on 66 kV and 11 kV lines considering a wide range of system parameters. Figure 4.7a shows the magnitude of current on both sides of the transformer when an external fault is created on the 11 kV line. The relevant differential and biased currents calculated using the MFCDFT method are shown in Fig. 4.7b.

It is to be observed that the differential current remains well below the biased current. Figure 4.7c shows the I_{diff}/I_{bias} trajectory which is almost at zero levels and hence, no trip signal is issued. Thus, the proposed scheme remains stable (inoperative) during any external fault condition.

4.5.5 External Fault Condition with CT Saturation

During a severe external fault condition, transformer protection faces problem as the CTs connected on both side observes the different level of saturation leading to

Table 4.1 Performance of the proposed algorithm during different types of internal faults

S. No.	Fault cases (Percentage winding from the terminal) (%)	Types of faults	Operating time (ms)
Without fault resistance (R_f = 0)			
1	5	Line-ground	31.2
2	10	Line-ground	31.2
3	15	Line-ground	31.23
4	50	Line-ground	31.41
5	90	Line-ground	32.08
6	95	Line-ground	32.27
7	50	Double-line-ground	31.06
8	50	Triple line	31.01
With fault resistance (R_f = 10 Ω)			
9	5	Line-ground	33.12
10	10	Line-ground	33.28
11	15	Line-ground	33.58
12	50	Line-ground	34.03
13	90	Line-ground	34.89
14	95	Line-ground	35.07
15	50	Double-line-ground	33.68

substantial differential current. Thus, the biased differential relay may operate during this situation if the saturation of CT is not detected in time. To check the feasibility of the proposed scheme, authors have simulated various test cases of the external fault including light to heavy CT saturation [22].

An adaptive criterion to avoid maloperation of the relaying scheme under CT saturation condition is applied as discussed in Sect. 4.4. An adaptive single slope biased differential characteristics are obtained in connection with the calculated level of CT saturation (Sect. 4.4). The normal setting of biased differential characteristic is 30% and depending on the level of saturation detected, it adaptively shifts up to 70%.

Light CT Saturation:

Normally during an external fault condition, if the fault current is 15–20 times higher than rated CT current, the change in core flux is limited to a certain value which leads to saturation of CT. A light saturation of CT is created by a slight increase in the burden resistance of CT (2 Ω).

The saturation effect is observed after some cycle from the inception of external fault (three-phase fault on 11 kV line at 10 km from the bus). Figure 4.8a shows the trajectory of I_{diff}/I_{bias} on considered biased differential characteristic (30%), during light CT saturation. It is noticed that during this test case, the magnitude of differential

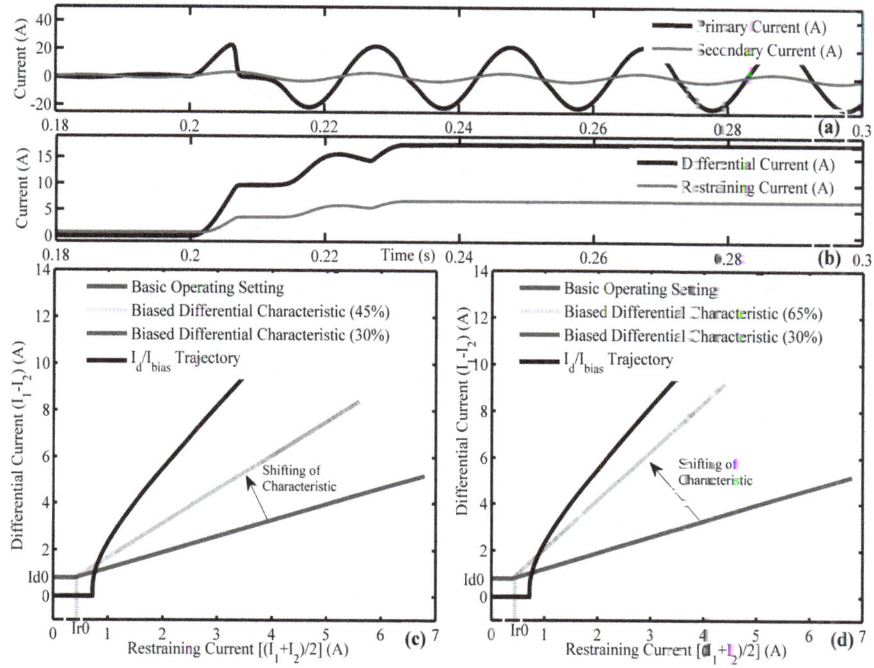

Fig. 4.6 Internal fault with CT saturation, **a** transformer primary and secondary current, **b** magnitude of differential and restraining current, **c**, **d** I_d/I_{bias} trajectory with medium and heavy CT saturation respectively

current slightly increases at the time of CT saturation, but remains lower than the biased current i.e. $I_{diff} < I_{bias}$ (Fig. 4.3). Thus, the relay does not operate and consider the test case as a pure external fault.

Medium CT Saturation

To analyze the medium CT saturation phenomenon, the burden resistance of CT secondary is moderately increased (6 Ω). In the event of an external fault (on 11 kV side), CT gets saturate within two-cycle and "I_{diff}" becomes greater than "I_{bias}" during the saturation period. At the same time, the proposed scheme detects this situation as a saturation condition, since δ_n becomes greater than the threshold (A_s) (Fig. 4.3). As a result, the biased differential characteristics adaptively change (as per Eq. 4.16) from 30 to 45%. Hence, the relay is blocked to issue trip signals during external fault with CT saturation. In this particular test case, it is observed that the value of δ_n is 20% higher than threshold A_s, thus,

$$\text{Degree of saturation} = \mathrm{X_s} = \left(1 - \frac{A_s}{\delta_n}\right) * 100\%$$

$$= 1 - \frac{1}{1.2} = 16.67\%$$

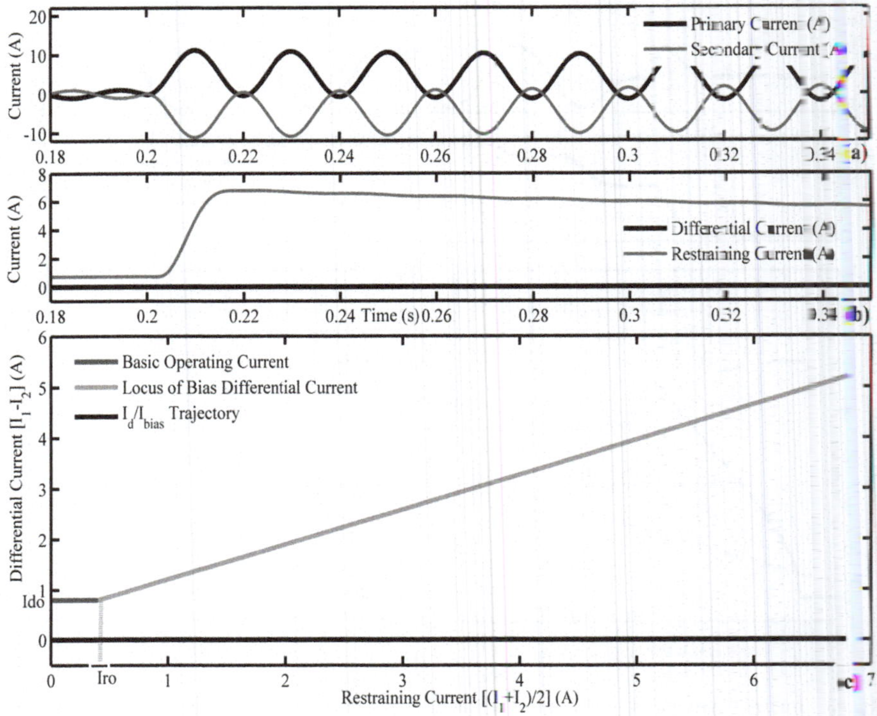

Fig. 4.7 External fault, **a** primary versus secondary current, **b** magnitude of differential and restraining current, **c** I_{diff}/I_{bias} trajectory

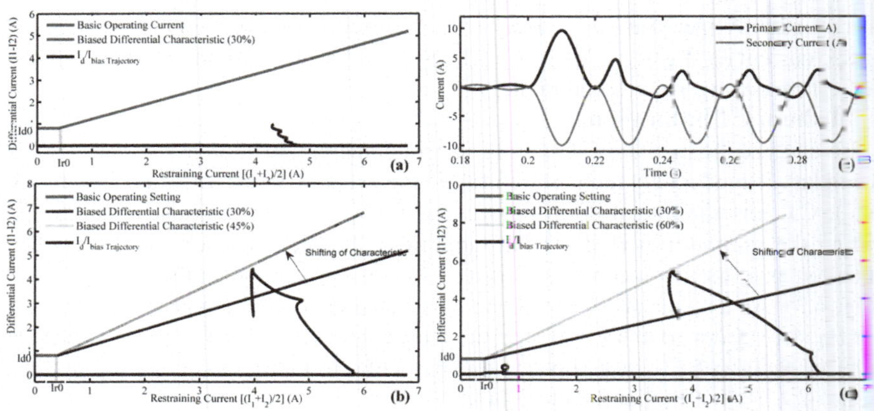

Fig. 4.8 I_{diff}/I_{bias} trajectory under various condition, **a** mild CT saturation, **b** medium CT saturation, **c** current during heavy CT saturation, **d** trajectory during heavy CT saturation

Hence, the slope to be revised with $X_s = 0.1667$ is

$$M_s = 0.3 + 0.9X_s = 0.3 + 0.9(0.1667) = 0.45$$

Here, 0.3 means 30% slopes as a basic setting, and 0.45 means 45% of the slope is required when a 20% saturation level is detected during an external fault. Figure 4.8b shows the trajectory of I_{diff}/I_{bias} and modified differential characteristics during medium CT saturation.

Heavy CT Saturation

To validate the proposed scheme, a severe CT saturation case is generated by increasing the secondary burden to 10 Ω during a close-in external fault (at 5 km on 66 kV line). The CT gets saturate within the first cycle from the inception of fault as shown in Fig. 4.8c. However, the algorithms effectively detect this condition and adaptively change the set characteristic to the new slope. Figure 4.8d shows the trajectory of I_{diff}/I_{bias} and purposefully shifting of characteristics from 30 to 60%. Thus, the possibility of maloperation of the biased differential relay is avoided during external fault with heavy CT saturation condition.

4.6 Comparison of the Studied Results with Traditional Solution

Comparative analysis with the existing scheme is carried out and the validation result is demonstrated here. It has been observed by the authors that the schemes based on sensitivity and security factors derived from operating and differential current [24] may not be able to identify CT saturation conditions. Hence, the above scheme may mal-operate in case of heavy saturation of CT, particularly during an external fault condition.

Conversely, the proposed algorithm provides accurate results from low to heavy CT saturation during both external faults. This fact can be easily understood by observing the comparative evaluation of the above scheme with the proposed scheme as shown in Fig. 4.9. Moreover, the proposed scheme operates during all internal faults even including CT saturation. Furthermore, the proposed scheme adaptively shifts the biased differential characteristic based on the level of saturation detected during an external fault. Thus, it provides better sensitivity during internal fault and stability in case of an external fault. It is to be noted from Fig. 4.9c that the adaptive characteristic of the existing scheme [24] is shifted to a higher slope as per the detection of the external fault within half cycle. However, due to the absence CT saturation detection facility in the algorithm, it mal-operates as the I_d/I_{bias} trajectory crosses the biased differential characteristic. On the other hand, the proposed scheme remains stable by identifying the level of deep CT saturation and adaptively shifting the biased characteristic as per requirement (70%).

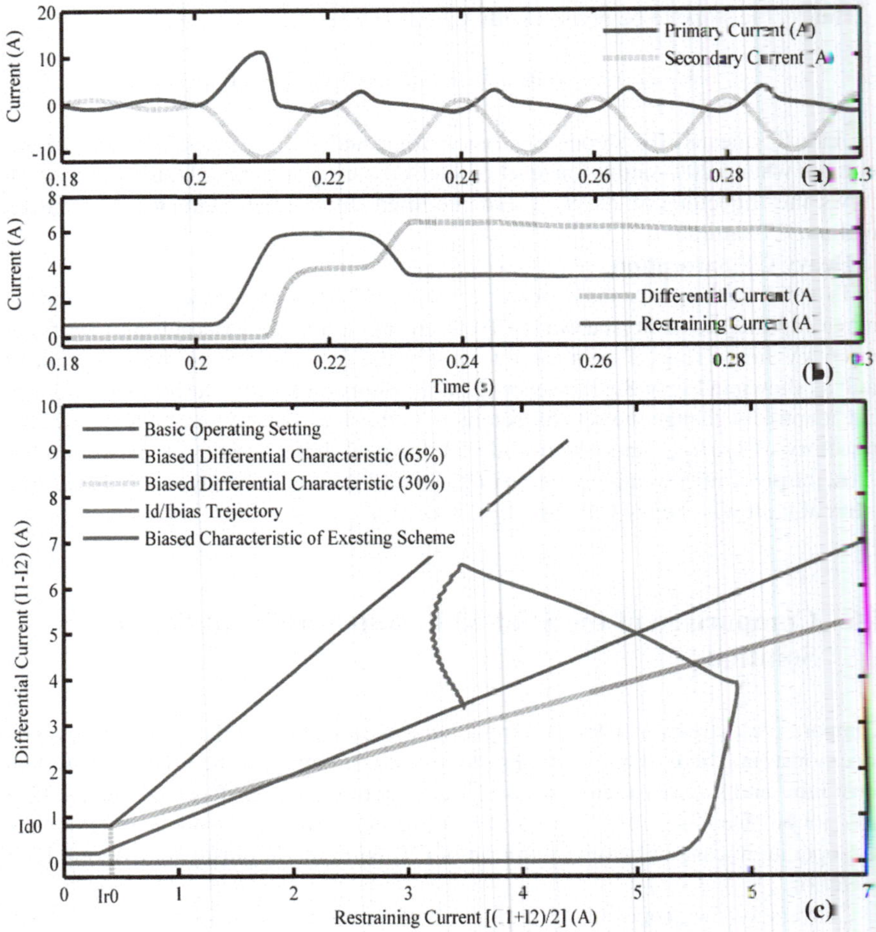

Fig. 4.9 External fault with heavy CT saturation, **a** transformer primary and secondary current, **b** magnitude of differential and restraining current, **c** I_d/I_{bias} trajectory with existing scheme [24] and proposed scheme

4.7 Hardware Implementation in Laboratory

Hardware is set up in the laboratory to validate the proposed adaptive algorithm on 2 kVA, 230/110 V multi tapping transformer. To simulate physical faults in the proposed hardware primary side 230 V are connected to electricity board supply and the secondary side is connected with lamp load. 5P10 protective classes CTs are connected in the primary and secondary side with 10/5 and 25/5 A rating respectively. Primary and secondary side internal faults and outside CT location external faults are generated through 12 A, 18 Ω variable resistors. Load and fault resistance is variable so adjustable fault current will be made as per the requirement. Additional 18 Ω, 12 A rheostat is connected on the secondary of the CT to commence saturation effect during internal as well as an external fault. Current sensors connected across the secondary side of primary and secondary CT to perform I to V conversion. The outputs of current sensors are given to high-resolution DSO to capture digital data. The setup parameter is given in Appendix.

Contactors are taken as a circuit breaker (CBs) and current sensor ACS712ELCTR-30A-T is used in hardware to scale down and sense the current in a secondary path of CT. Dedicated Digital Signal Controller (DSC), AVR ATmega 328P as computational hardware is employed in the present work for implementation of the protection scheme. ATmega 328P is also equipped with a large memory capacity of 2 K words of on-chip SARAM, 32 K words on-chip flash memory, and 64 K words off-chip SARAM memory that is sufficient to store large program [25]. The high-performance, 10-bit, 8 channels analog-to-digital converter (ADC) has a minimum conversion time of 500 ns. For the execution of an algorithm, code written in 'C' language using an embedded coder toolbox available in MATLAB is loaded in the memory of a processor. The communication between PC and DSC is done by programmable Universal Asynchronous Receiver Transmitter (UART) which is used to monitor the real-time measurements. A current sensor transfers the current signal into an equivalent 5-volt signal. Both the primary and secondary current sensor sends a signal to ATmega328 Microcontroller. In DSC there are facilities to download the I_d and I_{bias} trajectory results in .xls format. Chopping is provided in programming at 2 s for obtaining results acceptably.

4.7.1 Internal Fault Conditions

Figure 4.10a$_1$, b$_1$, c$_1$ shows the real-time data recorded using DSO, for the transformer primary and secondary currents, under internal fault without resistance, with saturated CT and during high resistance internal fault, respectively. Also, Fig. 4.10a$_2$, b$_2$, c$_2$ shows the I_d/I_{bias} trajectory and slope setting for the proposed algorithm. It is observed from Fig. 4.2 that for all internal faults the differential current always remains higher than biased current. Hence, it is solely authenticated that the algorithm operates under any type of internal fault with CT saturation and high resistance internal fault.

4.7.2 External Fault and Overload Condition

External fault without CT saturation and overload condition of the transformer are carried out to validate the proposed algorithm as shown in Fig. 4.11. The current signals captured by DSO from both sides of the transformer are shown in Fig. 4.11a$_1$, b$_1$. The trajectory of I_d/I_{bias} and the nominal slope setting of the differential relay are

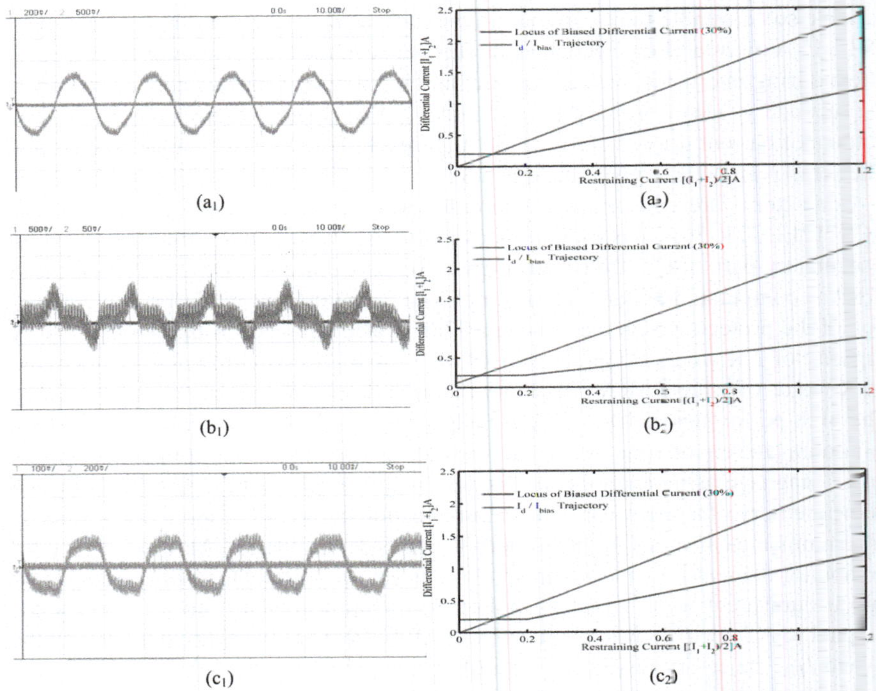

Fig. 4.10 **a**$_1$, **b**$_1$, **c**$_1$ Primary and secondary current waveform during internal fault and **a**$_2$ **b**$_2$, **c**$_2$ I_d/I_{bias} trajectory for internal fault with zero resistance, CT saturation under internal fault, high resistance internal fault

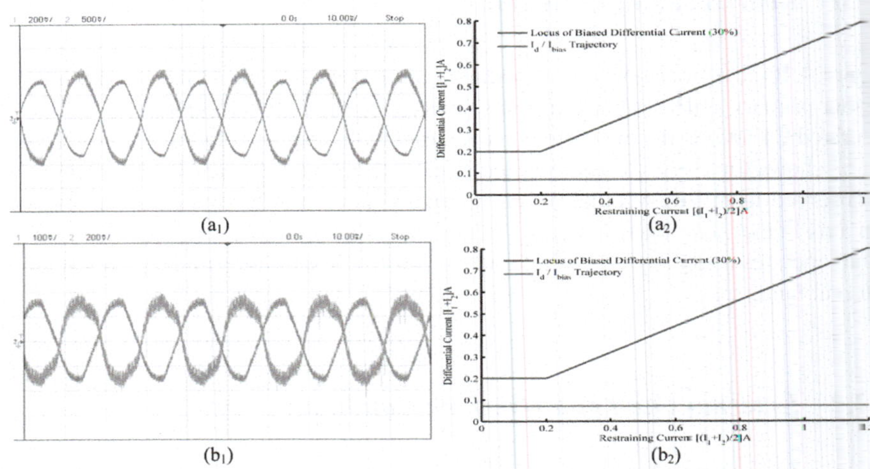

Fig. 4.11 **a**$_1$, **b**$_1$ Recorded primary and secondary current waveform and **a**$_2$, **b**$_2$ I_d/I_{bias} trajectory for external fault and overloading condition

shown in Fig. 4.11a$_2$, b$_2$. During external fault and 10% excess load on transformer secondary, I_d/I_{bias} trajectories remain well below the set value, hence relay does not issue trip signal.

4.7.3 External Fault with Light, Medium and Heavy CT Saturation Conditions

CT saturation under external faults is the main obstacle for protection engineers in the design of biased differential protection. During external fault with CT saturation, the relay may mal-operate and system reliability reduced. Three types of cases are considered for validation of the algorithm like light, medium, and heavy CT saturation. Figure 4.12a$_1$, b$_1$, c$_1$ shows recorded primary and secondary currents using

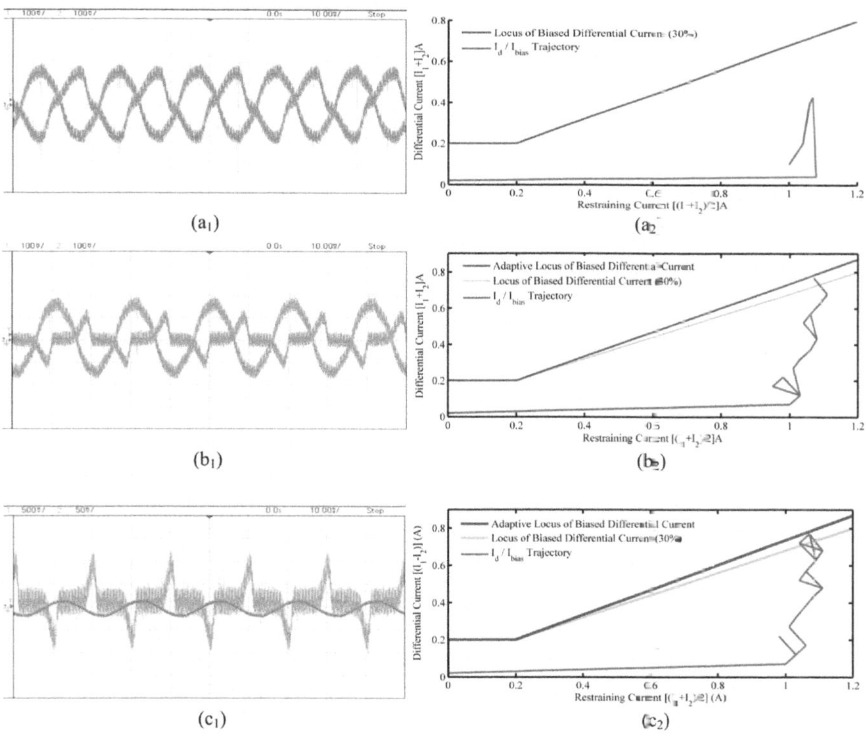

Fig. 4.12 **a**$_1$, **b**$_1$, **c**$_1$ External fault current waveform during low, medium and heavy CT saturation and **a**$_2$, **b**$_2$, **c**$_2$ I_d/I_{bias} trajectory for low, medium and heavy CT saturation under external fault conditions

DSO under external fault with light, medium, and heavy CT saturation conditions respectively. By increasing CT secondary burden (using variable resistance) all three conditions can be simulated.

Figure 4.12a$_2$ shows I_d/I_{bias} trajectory under light CT saturation during an external fault. But as per algorithm, it is not reached up to 30% locus of biased differential current as shown. It means the degree of saturation $D_{n(3)}$ is lesser than the threshold value of A_s, so as per Eq. 4.7, the degree of saturation is not reached up to set point. Hence adaptive shifting of slope setting is not required.

Figure 4.12b$_2$ shows the trajectory of I_d/I_{bias} during an external fault with medium CT saturation. It is to be noted that during this situation, the value of $D_{n(3)}$ is larger than the threshold value which shifts the slope adaptively to 36.9% as per Eq. 4.8. Hence, as per the developed logic, relay avoids mal-operation during external fault with medium CT saturation. For these experiment 185 Ω, 2 A rated variable resistor is placed in secondary of CT to generate saturation conditions. By increasing the resistance further in secondary of CT under external fault creates heavy CT saturation conditions. The adaptive shifting of differential characteristic slope and relay behavior during said heavy CT saturation is demonstrated in Fig. 4.12c$_2$. It is to be concluded that the developed relay remains stable (inoperative) under all types of external fault for light to heavy CT saturation condition.

4.7.4 Three Phase Transformer Hardware Results with Adaptive Shifting Characteristic Under CT Saturation Conditions

This algorithm is also validated on 50 kVA, 440/220 V transformer in a laboratory environment successfully (more information on detail of three phase transformer is depicted in topic Sect. 6.8 of Chap. 6). One of the results elaborated here under LL fault with one CT saturated condition. The proposed algorithm provides accurate results from low to heavy CT saturation during all external faults. Among them one of the test cases, one CT saturated under LL fault and I_d/I_{bias} trajectory shifting is elaborated here. Figure 4.13a shows three-phase DSO captured current waveforms under double line fault (LL) with one CT saturated. Under this situation, I_d/I_{bias} trajectory is shifted just as shown in Fig. 4.13b and system protected from mal operation under abnormal conditions like CT saturation under external fault.

The proposed algorithm provides accurate results from low to heavy CT saturation during both external faults. This fact can be easily understood by observing the evaluation of the above scheme as shown in Figs. 4.10, 4.11, 4.12 and 4.13 hardware results. Moreover, the proposed scheme operates during all internal faults even including CT saturation. Furthermore, the proposed scheme adaptively shifts the biased differential characteristic based on the level of saturation detected during an external fault. Thus, it provides better sensitivity during internal fault and stability in case of an external fault. It is to be noted that the adaptive characteristic of the

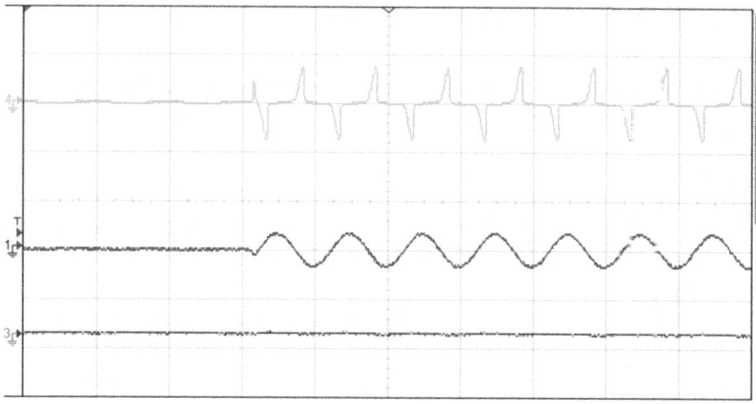

(a) L-L Fault with One CT Saturated

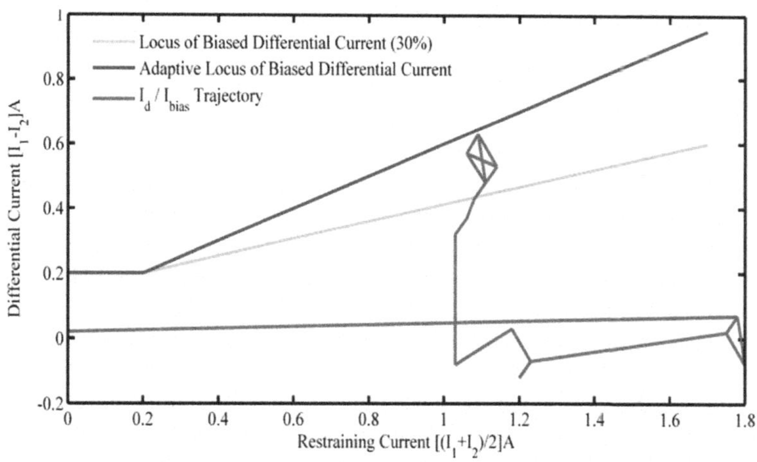

(b) I_d/I_{bias} Trajectory for External Fault with CT Saturation for L-L Fault

Fig. 4.13 Three phase hardware setup L-L fault (with one CT saturated) DSO results and shifting of adaptive percentage biased characteristics

existing scheme is shifted to a higher slope as per the detection of the external fault within half cycle. However, due to the absence CT saturation detection facility in the algorithm, it mal-operates as the I_d/I_{bias} trajectory crosses the biased differential characteristic. On the other hand, the proposed scheme remains stable (inoperative) by identifying the level of deep CT saturation and adaptively shifting the biased characteristic as per requirement (70%).

4.8 Novelty Introduced by the Proposed Scheme

Second harmonic ratio is used to detect inrush condition only and 3rd order derivative is a very accurate method to detect the saturation index. Other novelties of the schemes are as under,

(1) Extraction of fundamental frequency component using full-cycle discrete Fourier transform (FCDFT), which is extract signal with speed and accuracy.
(2) Adaptively change of biased differential characteristic slope concerning CT saturation.
(3) Use of third differential equation-based CT saturation detection which gives accurate detection concerning another method.
(4) The algorithm gives correct operation under heavy as well as light CT saturation in internal and external faults also.
(5) The algorithm is also validated in laboratory environments on a single phase and three phase transformers.

4.9 Summary

This paper presents a novel approach to adaptive protection of distribution transformer. The technique is based on the percentage biased differentials principle including a saturation detection method. Initially, it detects the magnetizing current based on the 2nd harmonics component derived using the FCDFT filter. The algorithm based on CT saturation evaluation and differential principle successfully discrimination between internal fault and external fault. The performance of the proposed algorithm is validated through several simulations, based on a 15 MVA, 66/11 kV distribution transformer, modeled in the PSCAD/EMTDC software environment. The algorithm is designed using MATLAB software to estimate differential & biased currents (using FCDFT), and third-order derivative of CT secondary current (saturation detection). The developed approach adaptively modifies differential relay characteristics during the saturation period of CTs. It is observed that the suggested scheme operates only during internal faults including a high resistance fault, and it remains inoperative during external faults and normal load conditions. Also, based on the comparative evaluation, the performance of the proposed scheme is found to be superior compare to the existing schemes. Moreover, the results indicate that the scheme considerably improves protection stability in cases of external faults during different CT saturation levels.

4.10 Published Article Based on This Work

- Dharmesh Patel, N. G. Chothani, K. D. Mistry, Dhaval Tailor, "Adaptive Algorithm for Distribution Transformer Protection to Improve Smart Grid Stability", DEGRUYTER, International Journal of Emerging Electric Power Systems, 19(7), pp. 1–14, 2018.

Appendix

Simulation model data:

Source data	3-phase, 20 MVA, 66 kV, 50 Hz
Line data	Length = 15 km, System voltage = 66 kV Positive-sequence impedance = 0.0297 + j0.332 Ω/km Zero-sequence impedance = 0.162 + j1.24 Ω/km Positive-sequence capacitance = 0.245 nF/km Zero-sequence capacitance = 0.375 nF/km
Transformer data	YY connected, 15 MVA, 66/11 kV, 3-phase, with 0.[illegible] pu leakage reactance (11 kV side of the transformer is connected to Distributed Generations)
CT data	Primary-350/2 A, Secondary-2100/2 A, Secondary winding resistance and inductance = 0.5 Ω and $0.8e^{-3}$ H
Load	P + jQ = (15 + 5j)

Equipment data for hardware:

Transformer data	2 KVA, 230/115 V, 1-phase, 50-Hz, %Z = 12
CT data	Primary side: 10/5 A, 15 VA, 5p10 and for secondary side 20/5 A, 15 VA, 5p10
Load	Lamp load, 25 A
Source data	1-phase, 0–300 V, 50 Hz, Variable supply from the electricity board

References

1. Abdelsalam HA, Abdelaziz AY (2015) A new strategy for selection of switching instant to reduce transformer inrush current in a single-phase grid-connected photovoltaic system. Electr Power Compon Syst 43(11):1297–1306
2. Hooshyar A, Sanaye-Pasand M (2012) Accurate measurement of fault currents contaminated with decaying DC offset and CT saturation. IEEE Trans Power Deliv 27(2):773–783

3. Samet H, Ghanbari T, Ahmadi M (2015) An auto-correlation function based technique for discrimination of internal fault and magnetizing inrush current in power transformer. Electr Power Compon Syst 43(4):399–411
4. Narayana GS, Badgujar KP, Kulkarni SV (2013) Factorisation-based transfer function estimation technique for deformation diagnostics of windings in transformers. IET Electr Power Appl 7(1):39–46
5. Valsan SP, Swarup KS (2008) Wavelet based transformer protection using high frequency power directional signals. Electr Power Syst Res 78(4):547–558
6. Bigdeli M, Vakilian M, Rahimpour E (2012) Transformer winding faults classification based on transfer function analysis by support vector machine. IET Electr Power Appl 6(5):268–276
7. Galdi V, Ippolito L, Piccolo A, Vaccaro A (2001) Parameter identification of power transformers thermal model via genetic algorithms. Electr Power Syst Res 60(2):107–113
8. Ahmadi M, Samet H, Ghanbari T (2015) Discrimination of internal fault from magnetising inrush current in power transformers based on sine-wave least-squares curve fitting method. IET Sci Meas Technol 9(1):73–84
9. Wei CH, Tang WH, Wu QH (2014) A hybrid least-square support vector machine approach to incipient fault detection for oil-immersed power transformer. Electr Power Compon Syst 42(5):453–463
10. Ozgonenel O (2013) Correction of saturated current from measurement current transformer. IET Electr Power Appl 7(7):580–585
11. Stanbury M, Djekic Z (2015) The impact of current-transformer saturation on transformer differential protection. IEEE Trans Power Deliv 30(3):1278–1287
12. Hajipour E, Vakilian M, Sanaye-Pasand M (2015) Current-transformer saturation compensation for transformer differential relays. IEEE Trans Power Deliv 30(5):2293–2302
13. Ajaei FB, Sanaye-Pasand M, Davarpanah M, Rezaei-Zare A, Iravani R (2011) Compensation of the current-transformer saturation effects for digital relays. IEEE Trans Power Deliv 26(4):2531–2540
14. Bak DJ, Dong XZ, Wang B, Shin SX, Rebizant W (2012) New method of detection of current transformer saturation. In: 11th IET international conference on developments in power systems protection (DPSP 2012, pp 1–5
15. Shi DY, Buse J, Wu QH, Guo CX (2013) Current transformer saturation compensation based on a partial nonlinear model. Electr Power Syst Res 97:34–40
16. Kang S, Kim M, Nam S, Shin JH, Jung J (2014) A CT saturation detection algorithm based on wavelet transformation. In: 12th IET international conference on developments in power system protection (DPSP 2014), pp 1–4
17. Esmail EM, Elkalashy NI, Kawady TA, Taalab AI, Lehtonen M (2015) Detection of partial saturation and waveform compensation of current transformers. IEEE Trans Power Deliv 30(3):1620–1622
18. Kang YC, Jin ES, Kang SH, Crossley PA (2004) Compensated-current differential relay for protection of transformers. IEE Proc Gener Transm Distrib 151(3):281–289
19. Wu QH, Lu Z, Ji T (2009) Protective relaying of power systems using mathematical morphology, 1st edn. Springer-Verlag, London Limited, London, New York
20. PSCAD Research Center (2005) EMTDC-transient analysis for PSCAD power system simulation. Winnipeg, MB, Canada
21. van Cortlandt Warrington AR (1968) Protective relays: their theory and practice, 2nd edn. Chapman & Hall, London
22. Chothani NG, Bhalja BR (2014) New algorithm for current transformer saturation detection and compensation based on derivatives of secondary currents and Newton's backward difference formulae. IET Gener Transm Distrib 8(5):841–850
23. Tan Q, Liu P, Miao S, Zhang W, Zhou L (2013) Self-adaptive transformer differential protection. IET Gener Transm Distrib 7(1):61–68
24. Zhang W, Tan Q, Miao S, Zhou L, Liu P (2013) Self-adaptive transformer differential protection. IET Gener Transm Distrib 7(1):61–68
25. Atmel (2016) ATmega328/P. AVR Microcontrollers, p 442

Chapter 5
Relevance Vector Machine Based Transformer Protection

This editorial presents a new scheme, based on Relevance Vector Machine (RVM) as a fault classifier. The developed algorithm is assessed by simulating various disorders on 345 MVA, 400/220 kV transformer in PSCAD/EMTDC™ software, and also on prototype model with 2 kVA, 230/110 V multi tapping transformer.

5.1 Literature Studied for the Idea Generation

The power transformer is one of the most vital and costly components in the power system. Due to the multifold growth of the power system network, a variety of abnormal conditions and faults can take place in it. Magnetizing inrush, internal and inter-turn faults are the critical types of conditions to detect within the transformer. There are various intelligent techniques available for discrimination of internal faults and other external abnormalities with minimum time.

Tripathy et al. [1] elaborated transformer protection using optimum Probabilistic Neural-Network (PNN) as a core classifier to detect a fault. However, classification efficiency is less. Balaga et al. [2] offered power transformer protection using a trained parallel hidden layered ANN-based Genetic Algorithm (GA) and tried to overcome pattern recognition error. However, in ANN, training, and testing are time-consuming and having fault detection process too complicated. Mittal et al. [3] proposed SVM based fault classification in transformer protection with Dissolved Gas Analysis (DGA) data collection. Bigdeli et al. [4] classify transformer fault based on analysis of transfer function with SVM and compared results with ANN techniques which prove better accuracy of SVM based techniques. Due to a large number of support vectors, SVM takes more time for classification. Now a day combination of SVM with ANN is utilized as a conventional technique in the research field. Koley et al. [5] proposed SVM and ANN-based transmission line protection under nonlinear load. In which, fault classification is based on SVM, and the location of the fault is defined by ANN.

D. Patel and N. Chothani, *Digital Protective Schemes for Power Transformer*, Power Systems, https://doi.org/10.1007/978-981-15-6763-6_5

Ashrafian et al. [6] described S-Transform based fault classification in power transformer. Gil and Abdoos [7] a proposed combination of S-transform and SVM based busbar protection schemes. Saleh et al. [8] offered a Wavelet Packet Transform (WPT) based transformer protection. However, the decomposition of the captured signal is a complex process and depends on the sampling rate. Chen et al. [9] offered high impedance fault detection with Daubechies db4 WT methodology. In contrast to phasor estimation, WT is more accurate for this reason. Medeiros et al. [10] proposed transformer protection using boundary discrete WT. Maya et al. [11] offered Empirical Wavelet Transform (EWT) to discriminate internal and external fault under various abnormal conditions. However, the Wavelet Transform based schemes require more attention to select various parameters such as wavelet type, level of decomposition, threshold, and other related parameters.

Compare to WT and ANN, a combination of SVM with Wavelet Packet Transform (WPT) [12, 13] is one of the good optimistic classifier methods for transformer protection. SVM based techniques are elaborated for long transmission line fault classification [12], whose parameters are optimized by Particle Swarm Optimization (PSO) with WPT. Shah et al. [13] proposed SVM based transformer protection, however during recovery inrush efficiency is around 92%. Recently Zhang et al. [14] elaborated discrimination of internal fault and magnetizing inrush conditions based on higher-order statistics and compared it with conventional second-order harmonic based restraining techniques which one is outdated now a day.

Relevance Vector Machine (RVM) has an analogous function with SVM with better simplification achievement and superior model discrimination ability which does not have to satisfy the Mercer's states. RVM is a sparse probability model proposed by M. E. Tipping based on the Bayesian learning theory [15]. Naveen et al. [16] proposed the application of relevance vector machines in real-time intrusion detection. Li [17] proposed the generalization performance of RVM by an incremental relevance vector machine algorithm. Lou et al. [18] elaborated reliability prediction with RVM software over various classifier techniques. Niu et al. [19] proposed an RVM application for transformer fault diagnosis using data mining technology. In comparison to the neural network and SVM, the proposed RVM technique has improved exceptional decision capability [20]. Therefore, implementation of the proposed RVM based transformer fault zone discrimination scheme is possible with recent signal processing techniques.

This research presents a new algorithm based on RVM for the classification of various faults and abnormalities in the transformer. It is observed that the proposed technique provides acceptable results, and can be used as a modern numerical relaying scheme for transformer protection. This method is compared with SVM and PNN classification considering wide variation in system and fault parameters. Section 5.2 describes system modelling and test data generation. Sections 5.3 and 5.4 elaborates on the proposed algorithm and methodology. Sections 5.5 and 5.6 depict simulation and hardware-based validation, respectively. Finally, Sect. 5.7 present advantage of the proposed scheme.

5.2 System Modeling and Data Generation

As shown in Fig. 5.1, Thevenin's equivalent generator is connected to 100 km, 400 kV line on the primary side of two numbers 345 MVA, 400/220 kV, YY Transformers and is connected to the infinite bus through 80 km, 220 kV transmission line of Kasor substation, Gujarat, India. The ratings of CTs are decided based on the rated current of the transformer considering 115% overload condition. Simulation of the existing Indian power system is done in PSCAD/EMTDC™ using user-developed modules [21]. To detect turn-to-turn faults in transformers, a model is developed with tapings taken out on primary and secondary of the transformer. An internal code in FORTRAN and dialog boxes is created to represent various devices in a graphical interface. The system parameter is given in Appendix.

Large numbers of fault cases and inrush situations have been created considering different types of fault (F_{type}), Source Impedance (SI), Fault Inception Angle (FIA), Fault Resistance (R_f), load angle (δ), and Fault Locations (FL) on transformer winding (F_{int}) as well as on bus/line (F_{ext}). Further, the Multi-Run block of PSCAD is used to alert the system parameter, to produce numerous simulation cases. Different parameter values that have been selected in this work are given in Tables 5.1 and 5.2 for internal and external faults, respectively with separation of training and testing data.

Table 5.1 shows 31,752 total data generated by various types of internal faults including winding faults, turn-to-turn faults, and interwinding faults. Among them, 16,875 data being considered as training, and remaining 14,877 data are chosen as validation of the proposed technique. Similarly, various external faults are generated on 400 and 220 kV bus including CT saturation. Also, external faults have been simulated at three different locations on both 400 kV and 220 kV transmission lines.

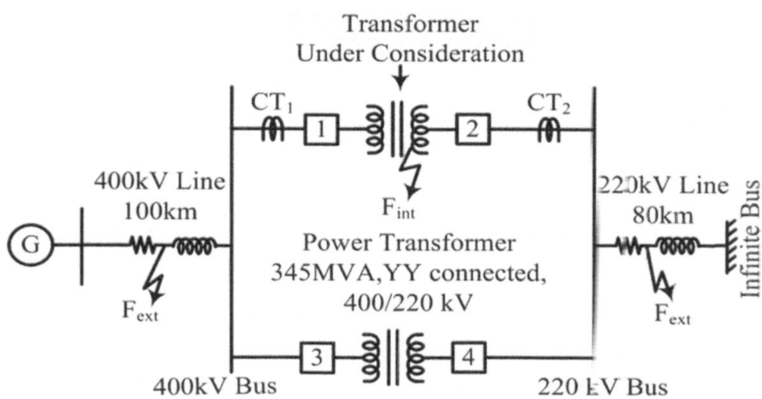

Fig. 5.1 Single line diagram for Indian power system

Table 5.1 Training and testing data considered for various internal faults

Parameter variation	Turn to turn fault	Turn to turn fault (Training data)	Primary to secondary winding fault	Primary to secondary winding fault (training)	Internal winding fault	Internal winding fault (training)
Fault type (F_{type})	**6** (3-on primary, 3-on secondary)	**6** (3-on primary, 3-on secondary)	**3** (all in three winding)	**3** (all in three winding)	**20** (10-primary + 10-secondary)	**20** (10-primary + 10-secondary)
Source impedance (SI)	**3** (75%, 100%, 125%)	**3** (75%, 100%, 125%)	**3** (75%, 100%, 125%)	**3** (75%, 100%, 125%)	**3** (75%, 100%, 125%)	**3** (75%, 100%, 125%)
Fault location (*FL*)	**6** (0.2%, 1%, 2%, 3%, 4%, 5%)	**5** (0.2%, 1%, 3%, 4%, 5%)	**6** (0%, 10%, 25%, 50%, 75%, 90%)	**5** (0%, 10%, 25%, 75%, 90%)	**6** (0%, 10%, 25%, 50%, 75%, 90%)	**5** (0%, 10%, 25%, 75%, 90%)
Fault inception angle (FIA)	**6** (0°, 30°, 60°, 90°, 120°, 150°)	**5** (0°, 30°, 90°, 120°, 150°)	**6** (0°, 30°, 60°, 90°, 120°, 150°)	**5** (0°, 30°, 90°, 120°, 150°)	**6** (0°, 30°, 60°, 90°, 120°, 150°)	**5** (0°, 30°, 90°, 120°, 150°)
Fault resistance (R_f)	–	–	**4** (0 Ω, 5 Ω, 10 Ω, 15 Ω)	**3** (0 Ω, 5 Ω, 15 Ω)	**4** (0 Ω, 5 Ω, 10 Ω, 15 Ω)	**3** (0 Ω, 5 Ω, 15 Ω)
Load angle (δ)	**3** (10°, 15°, 20°)	**3** (10°, 15°, 20°)	**3** (10°, 15°, 20°)	**3** (10°, 15°, 20°)	**3** (10°, 15°, 20°)	**3** (10°, 15°, 20°)
Total	1944	Training data = 1350, hence testing data = 594	3888	Training data = 2025, hence testing data = 1863	25,920	Training data = 13,500, hence testing data = 12,420

Hence, total training data = 16,875 and testing data = 14,877 for internal fault in the transformer
Bold represents a number of data/parameter used for the validation of the proposed scheme

Table 5.2 shows a total of 21,600 data simulated for various types of external faults. Among them, 13,500 cases are selected as training, and the remaining 8100 data are taken as testing of the proposed algorithm.

Inrush current is set up in the transformer when the primary of the transformer is being subjected to change in voltage keeping secondary in an open condition. Magnetizing inrushes are subdivided into three categories: (1) initial inrush (2) sympathetic inrush and (3) recovery inrush [22, 23].

Table 5.2 Training and testing data considered for various external faults

Parameter variation	Fault on 400 and 220 kV bus (with and without CT saturation)	Fault on 400 and 220 kV bus (with and without CT saturation) (training data)	Fault on 400 and 220 kV line	Fault on 400 and 220 kV line (training data)
Fault type (F_{type})	**20** ((L-g, LL-g, LL, LLL)10 * (two bus) 2)	**20** ((L-g, LL-g, LL, LLL)10 * (two bus) 2)	**20** ((L-g, LL-g, LL, LLL)10 * (two line) 2)	**20** ((L-g, LL-g, LL, LLL)10 * (two line) 2)
Source impedance (SI)	**3** (75%, 100%, 125%)	**3** (75, 100, 125%)	**3** (75%, 100%, 125%)	**3** (75%, 100%, 125%)
Fault location (FL)	**1**	**1**	**3** (1 km, 20 km, 50 km)	**3** (1 km, 20 km, 50 km)
Fault inception angle (FIA)	**6** (0°, 30°, 60°, 90°, 120°, 150°)	**5** (0°, 30°, 90°, 120°, 150°)	**6** (0°, 30°, 60°, 90°, 120°, 150°)	**5** (0°, 30°, 90°, 120°, 150°)
Fault resistance (R_f)	**4** (0 Ω, 5 Ω, 10 Ω, 15 Ω)	**3** (0 Ω, 5 Ω, 15 Ω)	**4** (0 Ω, 5 Ω, 10 Ω, 15 Ω)	**3** (0 Ω, 5 Ω, 15 Ω)
Load angle (δ)	**3** (10°, 15°, 20°)	**3** (10°, 15°, 20°)	**3** (10°, 15°, 20°)	**3** (10°, 15°, 20°)
Total	4320 * 2 = 8640	Hence training data = 5400 and testing data = 3240	12,960	Training data = 8100, hence testing data = 4860

Hence, total training data = 13,500 and testing data = 8100 for external fault outside transformer zone. Bold represents a number of data/parameter used for the validation of the proposed scheme

Inrush is normally generated during the no-load operation of the transformer and it is related to core saturation characteristics of transformers. Core saturation is depending on switching angle and so, magnetizing inrush current magnitude and peak values of positive or negative are defined as per the FIA. Hardware-based initial inrush analysis as shown in Fig. 5.1a. Researchers had tried to mitigate the inrush current which is generated at the time of transformer energization by refurbishing core material, bounding inception angle, etc.

Sympathetic inrush conditions are normally taking place during parallel operation of power transformers. When the 2nd transformer operated without load conditions, then that transformer itself getting inrush current and also affect in-service transformer is called sympathetic inrush. The dc component of the nearby transformer may lead to a saturating core of the in-service transformer. Sympathetic inrush is obtained on hardware-based analysis as shown in Fig. 5.1b.

When some phenomena take place in the power system like a sudden change in voltages followed by recovery of rated voltage due to synchronism in the power system, the effect of that sudden change and recovery will spread in the transformer operation. The sudden changes in voltages may appear because of fault clearance, voltage swings, momentary trip, auto reclosing, etc. The effect of the sudden changes is sound in the cases if the above-mentioned phenomena occur in the vicinity of the transformer. Though the effect is not that much prominent like initial inrush but may affect the transformer operation. The inrush generated when voltages recover to-rated synchronized voltage level is called recovery inrush condition. The effect of recovery inrush is dominant if the cleared fault is of three phases. Figure 5.1c shows the waveform of recovery inrush generated when a fault cleared nearby the transformer.

These all such inrush conditions are simulated under varying parameters as shown in Table 5.3.

It is to be noted from Table 5.3 that a total of 432 data is generated for various types of inrush conditions simulated in Fig. 5.1. It also shows the breakup of 315 training data and 117 testing data taken for validation of the proposed scheme.

For achieving better efficiency to identify internal transformer fault compared to external fault/abnormal conditions, proper training and testing data collections are highly important. Total 53,784 simulation cases as a whole have been considered out of which 30,690 fault cases (57.06% of 53,784) have been utilized for training of RVM whereas 23,094 fault cases (42.94% of 53,784) have been utilized for validation/testing of the proposed transformer protection technique as shown in Table 5.4.

Table 5.3 Training and testing data generated for various inrush conditions

Parameter variation	Initial inrush	Initial inrush (training data)	Sympathetic inrush	Sympathetic inrush (training data)	Recovery inrush	Recovery inrush (training data)
Source impedance (SI)	**3** (75%, 100%, 125%)	**3** (75%, 100%, 125%)	**3** (75%, 100%, 125%)	**3** (75%, 100%, 125%)	**3** (75%, 100%, 125%)	**3** (75%, 100%, 125%)
CB switching instant	**6** (0°, 30°, 60°, 90°, 120°, 150°)	**5** (0°, 30°, 90°, 120°, 150°)	**6** (0°, 30°, 60°, 90°, 120°, 150°)	**5** (0°, 30°, 90°, 120°, 150°)	**6** (0°, 30°, 60°, 90°, 120°, 150°)	**5** (0°, 30°, 90°, 120°, 150°)
Load angle (δ)	**3** (10°, 15°, 20°)	**3** (10°, 15°, 20°)	**3** (10°, 15°, 20°)	**3** (10°, 15°, 20°)	**3** (10°,15°, 20°)	**3** (10°, 15°, 20°)
Residual flux	**6** (0%, 10%, 25%, 45%, 60% and 80%)	**5** (0%, 10%, 45%, 60% and 80%)	–	–	–	
Total	324	Training data = 225, hence testing data = 99	54	Training data = 45, hence testing data = 9	54	Training data = 45, hence testing data = 9

Hence, total training data = 315 and testing data = 117 for magnetic inrush fault in the transformer. Bold represents a number of data/parameter used for the validation of the proposed scheme

Table 5.4 Total training and testing data collection for various conditions

Cases	Training data	Testing data	Total data
Internal fault	16,875	14,877	31,752
External fault	10,800	6480	17,280
External fault with CT saturation	2700	1620	4320
Inrush	315	117	432
Total	**30,690**	**23,094**	**53,784**

Bold represents a number of data/parameter used for the validation of the proposed scheme

Moreover, with training sets larger than 57.06%, the improvement in test error isn't found to be much significant. Hence, authors have selected 30,690 data out of 53,784 as a training set for RVM, which gives the best performance of 99.80% with 17 RVs.

Table 5.5 shows the structure of the comprehensive feature vector (empty matrix) which is formed by utilizing the training datasets separated from the total datasets and it has been used to train the RVM classifier.

5.3 Proposed Transformer Fault Classification Methodology

The fault zone prediction is formulated as a binary classification problem to determine whether the transformer internal fault or external fault/abnormal conditions. To begin, let vector $\delta \in R^n$ denote a pattern to be classified, and let scalar t denote its class label (i.e. $\mathrm{t} \in \tau \in 0, 1$). Also, let $(\delta_\mathrm{i}, \tau_\mathrm{i}), i = 1, 2, \ldots, N$ $(\mathrm{x_i}, \mathrm{t_i}), \mathrm{i} = 1, 2, \ldots, \mathrm{N}$ denote a given set of N training examples where each sample $\delta_\mathrm{i}\,\mathrm{x_i}$ has a known class label τ_i, t_i so, a classifier $f(\delta)\mathrm{f(x)}$ can correctly classify an input pattern.

5.3.1 RVM Classifier Model

For, given input vector δ, RVM classifier models the probability sharing of its class labels $\tau \in 0, 1\,\mathrm{t} \in$ using a sigmoid logistic function ρ as [18]:

$$\rho(\tau = 1|\delta) = \frac{1}{1 + \exp(-f_{RVM}(\delta))} \tag{5.1}$$

$$\mathrm{p(t = 1/x)} = \frac{1}{1 + \exp(-\mathrm{f_{RVM}(x)})}$$

where, $f_{RVM}(\delta) f_{RVM}(x)$ is the classifier, known as per Eq. 4.2

Table 5.5 Empty feature vector for training datasets

Simulation cases (training)	Current samples of phase R (80 samples/cycle * 2 side (primary and secondary))						Current samples of phase Y (80 samples/cycle * 2 side (primary and secondary))						Current samples of phase B (80 samples/cycle * 2 side (primary and secondary))					
	1	2	–	–	79	80 * 2	1	2	–	–	79	80 * 2	1	2	–	–	79	80 * 2
Case-1																		
Case-2																		
Case-3																		
Case-4																		
||																		
||																		
Case-n																		

Bold represents a number of data/parameter used for the validation of the proposed scheme

$$f_{RVM}(\delta) = y(\delta; \varpi) = \sum_{i=1}^{N} \varpi_i k(\delta, \delta_i) + \varpi_0 = \theta\, \varpi \tag{5.2}$$

where, N is the length of the data, weight vector $w = [w_0, , w_N]^T$ and θ is the $N \times (N+1)$ design matrix with $\theta = [\theta(\delta_1), \theta(\delta_2), \ldots, \theta(\delta_N)]^T$

$$\Phi = [\phi(x_1), \phi(x_2), , \phi(x_N)]^T$$

wherein $\phi(x_n) = [1, K(x_n, x_1), K(x_n, x_N)]^T$ and $K(x, x_i)$ is a kernel function.

Adopting the Bernoulli distribution for $\mathrm{P}(\tau/\delta)$, $p(t/x)$, the likehocd is given by:

$$\mathrm{P}(\tau|\varpi) = \prod_{n=1}^{N} \rho\{\lambda(\delta_n; \varpi)\}^{\tau_n}[1 - \rho\{\lambda(\delta_n; \varpi)\}]^{1-\tau_n} \tag{5.3}$$

where the target vector $\tau = [\tau_1, \ldots, \tau_N]^T$ with the targets $\tau_n \in \{0, 1\}$. A zero-mean Gaussian prior distribution over ϖ with variance χ^{-1} is added as:

$$p(\varpi|\chi) = \prod_{i=0}^{N} N\left(\varpi_i|0, \chi_i^{-1}\right) = \prod_{i=0}^{N} \sqrt{\frac{\chi_i}{2\pi}} \exp\left(-\frac{\chi_i}{2}\varpi_i^2\right) \tag{5.4}$$

where hyperparameter $\chi = [\chi_0, \chi_1, \ldots, \chi_N]^T$.

An individual hyperparameter associates independently with every weight. The posterior distribution over the weight from Bayes rule is thus given by:

$$\mathrm{P}(\varpi|\tau, \chi) = \frac{Likelihood \times prior}{Normalizing\ factor} = \frac{\mathrm{P}(\tau|\varpi)\mathrm{P}(\varpi|\chi)}{\mathrm{P}(\tau|\chi)} \tag{5.5}$$

Contrasting the regression case, still, the marginal likehood $\mathrm{P}(\tau|\chi)$ can no longer be obtained analytically by integrating the weights because of the discontinuity of the likelihood $\mathrm{P}(\tau|\chi)$, and an iterative method has to be used.

Let χ_i^* denotes the greatest a posteriori (MAP) approximation of the hyperparameter χ_i. The MAP estimate for the weights, denoted by ϖ_{MAP}, can be obtained by maximizing the posterior distribution of the class labels given the input vectors. This is comparable to maximizing the following objective task:

$$\mathrm{Z}(\varpi_1, \varpi_2, \ldots, \varpi_N) = \sum_{i=1}^{N} \log\mathrm{P}(\tau_i|\varpi_i) + \sum_{i=1}^{N} \log\mathrm{P}\left(\varpi_i\ \chi_i^*\right) \tag{5.6}$$

where the first summation term corresponds to the likehood of the class labels, and the second term corresponds to the prior on the parameters w_i. In the resulting solution, only those samples associated with nonzero coefficients ϖ_i (called relevance vectors) will contribute to the decision function.

The gradient of the objective function J concerning ω is:

$$\nabla Z = -X\varpi - \vartheta^T(\Gamma - \tau) \tag{5.7}$$

where, $X = (\chi_0, \chi_1, \ldots, \chi_N), \Gamma = [\rho(\lambda(\delta_1)), \rho(\lambda(\delta_N))]^T$, the matrix θ has elements $\theta_{i,j} = k\left(\delta_i, \delta_j\right)$. The Hessian of J is:

$$H = \nabla^2(Z) = -\left(\vartheta^T \Upsilon \theta + X\right) \tag{5.8}$$

where, $Y = (y_1, \ldots, y_N)$ is a diagonal matrix with $y_i = \rho(\lambda(\delta_i))[1 - \rho(\lambda(\delta_i))]$.

The posterior is approximated around ϖ_{MAP} by a Gaussian approximation with covariance:

$$\psi = -(H|\varpi_{MAP})^{-1} \tag{5.9}$$

And mean:

$$\mu = \varpi_{MAP} = \psi * \theta^T y\tau \tag{5.10}$$

Using the new ϖ_{MAP} the new target τ^* is then obtained through:

$$\tau^* = \theta \cdot \varpi_{MAP} + \Upsilon^{-1}(\tau - \rho(\lambda(\delta_n; \varpi))) \tag{5.11}$$

Using ψ and ϖ_{MAP}, hyperparameter χ_i can be updated by:

$$\chi_i = \frac{\gamma_i}{\varpi_{MAP}^2}, \tag{5.12}$$

where χ_i is the ith posterior mean weight and we have defined the quantities by:

$$\gamma_i = 1 - \chi_i * \psi_{ii}$$

With ψ_{ii} and the ith diagonal element of the posterior weight covariance, the regularization parameters of RVM are computed [18]. Once, the optimal regularization parameters 'ρ' and 'ψ_0' are estimated by learning procedure, the decision boundary of RVM is set for better data classification accuracy.

5.3.2 SVM Learning Model

SVM classifier maps the input data vector δ into a higher dimensional space F through an underlying nonlinear mapping $\theta(\delta)$ and then applies linear classification in this mapped space. Introducing a kernel function $k(\delta, \lambda) = \theta(\delta)^T\theta(\lambda)$, the SVM classifier $f_{SVM}(\delta)$ is given as under:

$$f_{SVM}(\delta) = \sum_{i=1}^{N} \varpi_i \, k(\delta, s_i) + b \tag{5.13}$$

where, $s_i, i = 1, 2, \ldots, N_s$, is a subset of the training samples $\{\delta_i, i = 1, 2, \ldots, N\}$ (which are called support vectors).

5.4 Proposed RVM Based Algorithm

Figure 5.2 shows a schematic block diagram of the proposed RVM based transformer fault classification algorithm. Samples of current signals of CT_1 and CT_2 located on primary and secondary of the transformer are acquired by the data acquisition system. The training of the RVM model is done by the Probabilistic Bayesian Learning (PBL) algorithm developed in MATLAB for accurate estimation of results at low computational time and cost. The training of RVM is done offline. Once the RVM is trained, the trained model developed is used online for the classification of real-time fault in the power system.

To train the RVM using PBL, 30,690 fault cases which are 57.06% of total 53,784 cases have been considered. While training the RVM, regularization parameters must be determined for a particular kind of kernel function. This gives the final form of the decision function [i.e. RVs and its associated coefficients ϖ_i as given by Eq. 5.2]. It is to be observed that the RBF kernel offers the least error value of 1.02% for the RVM classifier and the lowest error value of 3.89% for the SVM classifier. It is observed that only 17 numbers of Relevance Vectors (RVs) are produced during RVM training compare to the number of SVs found to be 159 in the SVM classifier. The RVM classifier is much sparser than the SVM thus; it provides accurate results with minimum time. Afterward, the trained model of RVM is further utilized for validation of the proposed algorithm using the test data set.

After configuring the RVM, the fault detection algorithm discriminates between the fault condition and the normal condition of the power system [24]. All the conventional digital/numerical relays detect fault/abnormal conditions in the very first stage of its performance. The proposed algorithm has two independent data acquisition paths, one for a fault detector unit and the other for the fault zone identification unit. At any time the fault is detected, samples of one cycle post fault current (80 samples) of both CTs are combined to formulate different feature vectors.

Each fault simulation case generates a feature vector of 480 samples consisting of 2 CT_s (as shown in Fig. 5.1) × 3 phases × 80 samples per cycle. Hence, with the help of testing data set for the cases mentioned in Table 5.4, a simulation database of *test data length* × *post fault samples* (i.e. 23,094 × 480) is generated. Soon after, these feature vectors are used as an input to trained RVM classifier. The output of the RVM ('+1' denotes internal fault and '−1' denotes external fault or inrush condition) is used to identify the fault condition/zone. The SVM classifier is also trained off-line

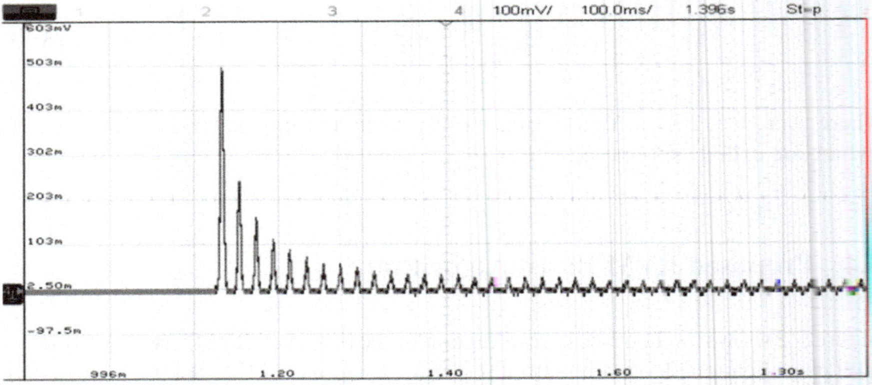

(a) Initial Inrush

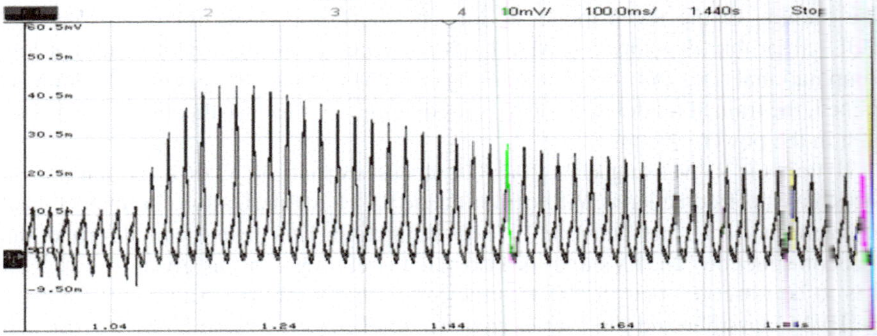

(b) Sympathetic Inrush

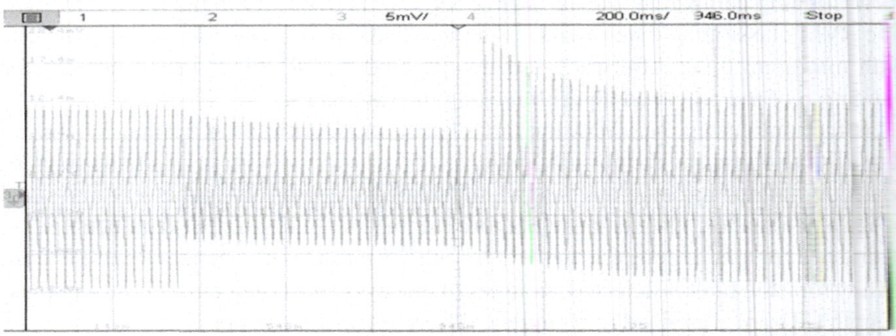

(c) Recovery Inrush

Fig. 5.2 Types of inrush

and tested similarly to that of the RVM classifier as described in Fig. 5.2. The fault classification accuracy is given by Eq. 5.14.

$$\eta\ \% = \frac{Correct\ Fault\ Classification}{Total\ Numbers\ of\ Test\ Data\ (23{,}094)} * 100\% \tag{5.14}$$

5.5 Result Analysis and Discussion

In the proposed scheme, fault cases classified correctly are represented as True Positive (TP) whereas fault cases categorized indecently are considered as True Negative (TN). Validation of the proposed technique is done on 23,094 test cases and the results in terms of classification accuracy are described in Tables 5.6, 5.7 and 5.8.

Table 5.6 shows classification accuracy for different fault cases. It is to be noted that, out of total test data, 23,047 are truly positive and 47 are true negative. So classification efficiency of RVM based technique is 99.8%. Moreover, internal faults are also subdivided into three parts turn to turn, primary to secondary, and internal winding faults under various parameter considerations. The accuracy of 99.69% during all internal fault shows the faithfulness of the proposed RVM based algorithm. Moreover, during various inrush conditions and external fault conditions, the scheme provides more than 99% fault classification accuracy. This indicated that the scheme

Table 5.6 Classification accuracy for different fault cases

Sr. No.	Faults cases	Faults/abnormalities	Numbers of test cases	TP	TN	Efficiency (%)
1	All types of internal faults	Turn to turn	594	590	4	99.32
		Primary to secondary winding	1863	1846	17	99.08
		Internal winding	12,420	12,396	24	99.80
2	External faults	400 kV bus	810	810	00	100
		220 kV bus	810	810	00	100
		400 kV line	2430	2429	01	99.95
		220 kV line	2430	2430	00	100
3	Inrush conditions	Initial inrush	99	99	00	100
		Sympathetic inrush	9	9	00	100
		Recovery inrush	9	9	00	100
4	External fault with CT saturation		1620	1619	01	99.69
	Total data		**23,094**	**23,047**	**47**	**99.80**

Bold represents a number of data/parameter used for the validation of the proposed scheme

Table 5.7 Fault type wise classification accuracy

Sr. No.	Fault type	Internal winding fault				External fault with and without CT saturation			
		Total	TP	TN	Efficiency (%)	Total	TP	TN	Efficiency (%)
1	L-g	3726	3717	9	99.75	2430	2429	01	99.95
2	L-L	3726	3720	6	99.83	2430	2429	01	99.95
3	L-L-g	3726	3717	9	99.75	2430	2430	00	100
4	L-L-L	1242	1242	00	100	810	810	00	100
	Total	**12,420**	**12,396**	**24**	**99.80**	**8100**	**8098**	**02**	**99.97**

Bold represents a number of data/parameter used for the validation of the proposed scheme

is more robust for transformer fault identification and at the same time it remains stable for all external disturbances. Also, the proposed algorithm can detect high resistance internal fault and successfully distinguishes external fault with severe CT saturation.

Figure 5.3 shows the current signal waveform of the various abnormalities simulated on the considered power system (Fig. 5.1). Considering 4 kHz sampling frequency, all the events are recorded for one cycle post-disturbance starting from 0.2 s to form the feature vectors.

The correctness of the proposed RVM based scheme in terms of identification of the different types of faults is depicted in Table 5.7. Considerations of various parameter settings are shown in Tables 5.1, 5.2 and 5.3. Total 20,520 test data are considered for validation of algorithm which is divided into an internal fault (12,420) and the external fault with and without CT saturation conditions (8100). It is to be observed that the proposed algorithm gives more than 99% accuracy in all ten types of fault simulated on the power system. (Here, ten types of faults mean 3-Line to ground fault, 3-line to line fault, 3-line-line to ground fault and 1-triple line fault. Means a total of 10 types of fault.)

Table 5.8 shows a comparison of proposed RVM, SVM, and PNN based classifier technique with a total of 23,094 test data. It is to be noted from Table 5.8 that the proposed RVM based scheme is intelligent to provide effective discrimination. The fault zone identification accuracy given by proposed RVM, SVM [13] and PNN [1] classifier during internal faults are 99.69%, 98.67% and 97.70%, respectively. At the same time, the proposed RVM based scheme offers improved stability during external faults and inrush condition as it gives an accuracy of the order of 100% compared to SVM and PNN based schemes. The proposed technique achieves overall classification efficiency of 99.79% compared to 98.77% of SVM and 97.94% of PNN based technique. It is also observed that the offline training time [25] of RVM (37.45 s) based technique is higher than SVM (64.27 s) and PNN (72.80 s) based scheme. On the other hand, the online testing time (11.72 s) of the proposed scheme is lesser than the SVM scheme (32.08 s) and PNN scheme (27.64 s) for total data classification.

Table 5.8 Comparisons of the proposed RVM Scheme with SVM and PNN scheme

Sr. No.	Faults cases/abnormalities	Total test cases	Proposed schemes (RVM)			SVM scheme			PNN scheme		
			TP	TN	%η	TP	TN	%η	TP	TN	%η
1	Internal faults	14,877	14,832	45	99.69	14,680	197	**98.67**	14,535	342	**97.70**
2	External faults	6480	6479	01	99.98	6419	61	**99.05**	6394	86	**98.67**
3	Inrush conditions	117	117	00	100	114	03	**97.43**	112	05	95.72
4	An external fault with CT saturation	1620	1619	01	99.93	1597	23	**98.58**	1578	42	**97.40**
		23,094	**23,047**	**47**	**99.79**	**22,810**	**284**	**98.77**	**22,619**	**475**	**97.94**

Bold represents a number of data/parameter used for the validation of the proposed scheme

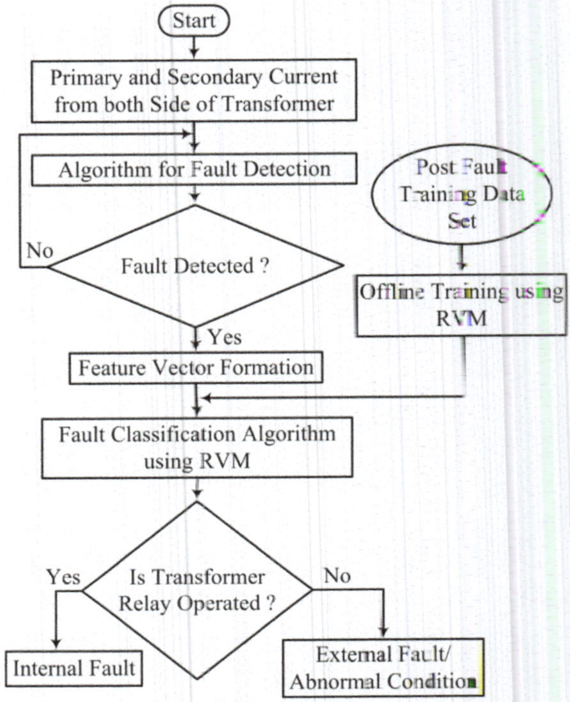

Fig. 5.3 Proposed RVM based fault classification algorithm

This indicates that the proposed scheme is faster and efficient in decision making than the SVM and PNN based scheme in the practical field.

Moreover, it does not require any preprocessing of the current date and multifold cross-validation as needed in SVM and ANN-based techniques.

Figure 5.4 represents the pattern of feature vector given as an input to the RVM classifier for different test conditions. It is to be observed from Fig. 5.4 that the feature vectors of transformer internal fault cases reasonably diversify with external abnormalities (Fig. 5.5).

5.6 Hardware Setup and Test Results

Figure 5.6 shows laboratory hardware setup for transformer protection and to simulate physical faults. In the proposed hardware as per the schematic diagram shown in Fig. 5.6a, primary side 230 volts are connected to electricity board supply and the secondary side is connected with lamp load. Two CTs are connected in the primary and secondary side with 10/5 and 25/5 A ratings respectively. Primary and secondary side internal faults are generated through S_1 and S_2 switch respectively which is connected through 12 A, 18 Ω variable resistors. External faults are created

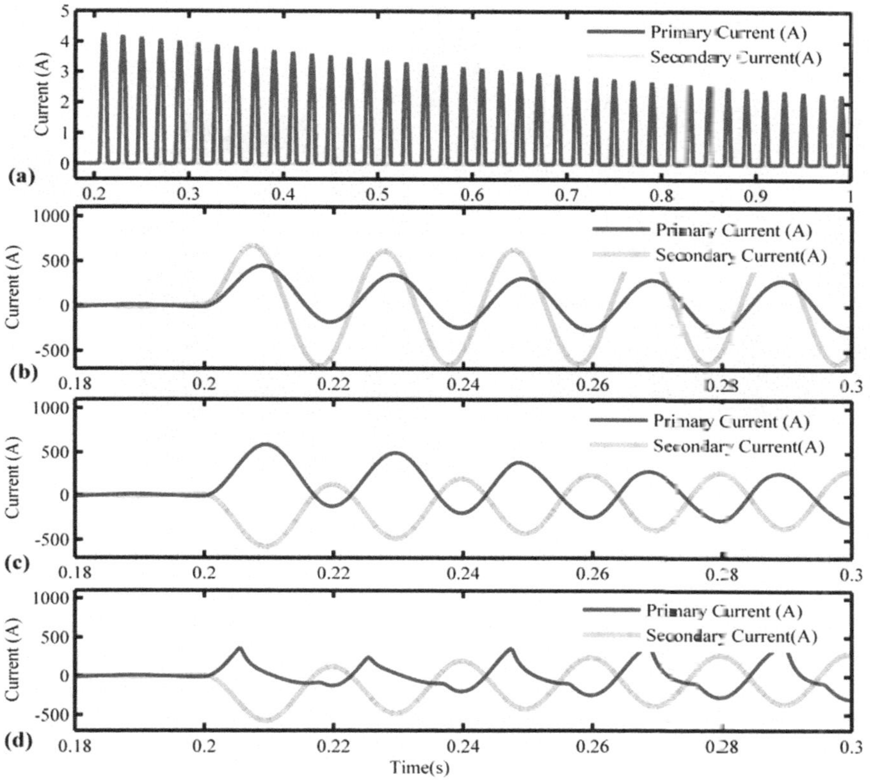

Fig. 5.4 Primary and secondary current waveform under **a** inrush condition **b** internal fault **c** external fault and **d** CT saturation condition

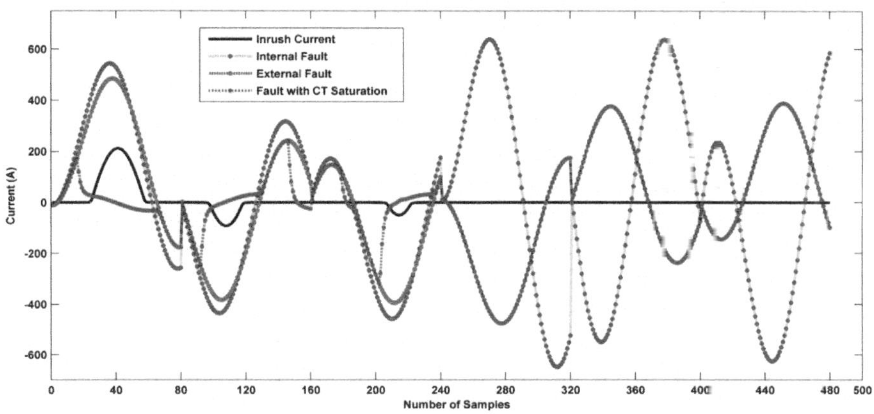

Fig. 5.5 Current signals during different fault/inrush conditions

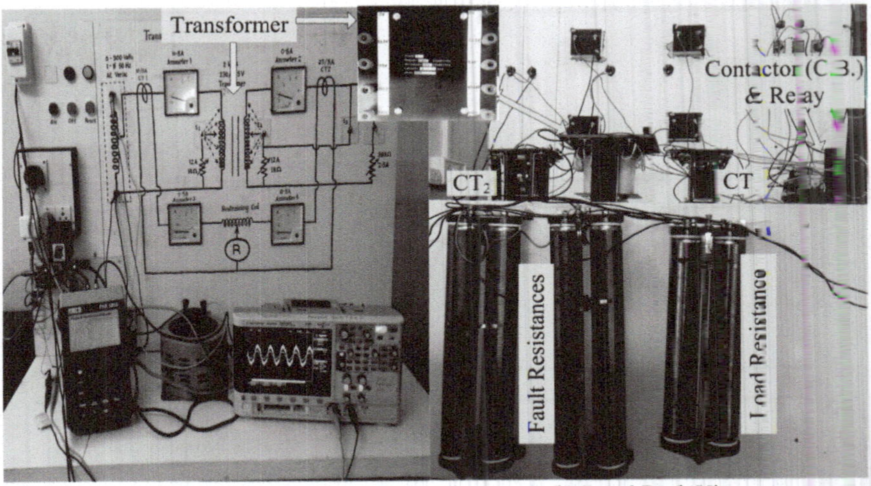

(a) Panel Front View (b) Panel Back View

Fig. 5.6 Hardware setup in the laboratory for transformer fault analysis

through switch S_3 and load is connected through switch S_4 as shown in Fig. 5.6a. Load and fault resistance is variable so adjustable fault current will be made as per the requirement. Additional 18 Ω, 12 A rheostat is connected on the secondary of the CT to commence saturation effect during internal as well as an external fault. It is to be noted that the proposed RVM algorithm is validated for all internal fault cases and issues trip signal on the output port within 20–22 ms.

As shown in Fig. 5.6a, faults are simulated using fault selector switch S_1 and S_2 respectively for internal and external. Multi-terminal Transformer is designed in such a way so the internal and inter-turn fault can easily be generated via selector switch at a different percentage of winding. The primary side and secondary side of the transformer are divided inappropriate four different voltage levels. Primary side is divided in 0–53.5–115–199.8–230 V and secondary side is divided 0–25.5–55–95.48–110 V. Various systems and fault parameters are varied to simulate the different faults. To evaluate the performance of the proposed algorithm during heavy CT saturation conditions, protective class CT is used. R_{ct1} and R_{ct2} are variable resistors inserted in secondary CTs as a burden to generate the effect of saturation CT_1 and CT_2. As shown in Fig. 5.6a, CS_1, and CS_2 are current sensors connected across the secondary side of CT_1 and CT_2 to perform I to V convertor. The outputs of CS_1 and CS_2 are given to high-resolution DSO to capture digital data. The collected data in digital form (sampled at 4 kHz) are migrated to the computer to train and test the RVM algorithm. As per the classification of an internal and external fault condition, the algorithm provides an output at the serial port of the computer. The output of RVM ('+1' denotes internal fault and '−1' denotes external fault/inrush conditions) is used to activate the main relay ('R') contact as shown in Fig. 5.6b. Figure 5.6b shows the control circuit implemented in the laboratory for real-time protection of a transformer. Contactor 'C' works as a circuit breaker (CB) coil. During normal operation of the transformer, when 'C' energizes by pressing spring-loaded 'ON' push button, two of its contacts, 'C_1' and 'C_2', as shown in Fig. 5.6a close to giving supply to the transformer. One of its contacts 'C_3' provides a hold-on path for continuous energization of the breaker coil. For manual de-energization of CB and to disconnect the transformer from the supply, press the 'OFF' push button provided in series with the contactor coil.

In the event of any internal fault in the transformer, the proposed RVM algorithm issues a trip signal ('+1') to the control circuit (Fig. 5.6b) at relay contact 'R', it energizes the auxiliary relay (AX). Simultaneously, one of its contacts 'AX_1' connected in series with coil 'C' open out. As a result, this coil C de-energizes, and thus all its contacts (as shown in Fig. 5.6a) is now opened out to disconnect the supply of the transformer. Another contact 'AX_2' of auxiliary relay provides hold on a path for continuous energization of 'AX'. After acknowledging the transformer internal fault, one has to press the "reset" puss button to de-energize the 'AX' relay.

Figure 5.7 illustrates the waveform captured by a high-resolution digital storage oscilloscope (DSO) for various fault events. Table 5.9 demonstrate the data generated by simulating various fault conditions on the hardware setup in the laboratory. This table also provides the separation of training and testing cases.

Total 60 numbers of physical fault data (6-inrush + 27 internal fault + 27 external faults) generated on the hardware setup. Out of this, 40 data are used for training, and the remaining 20 data are used for testing (Fig. 5.8).

It is to be observed that the fault classification accuracy obtained by RVM and SVM algorithm is 100%. Whereas, the fault classification accuracy gained by PNN scheme is 95% during validation using laboratory-generated fault data. Here, PNN misclassifies one test case of an external fault with CT saturation during validation. As the accuracy of PNN is lower compare to RVM and SVM for real-time data classification, it is not reliable for real-time protection of power transformer. On the other hand, due to fewer RVs compare to SVs, the execution time of the RVM algorithm is less than SVM. Hence, the proposed RVM algorithm provides fast fault discrimination for real field data. It is to be stated that while validating the algorithm with practical inrush and fault data, the signature of current signal produces for all internal fault widely differ with the pattern of current signal produce for all external fault.

5.7 Advantages of the Proposed RVM Based Scheme

Complete analysis of the simulations and hardware results presented in previous sections based on the proposed RVM technique emphasizes different advantages over the existing schemes and they are recapped as below.

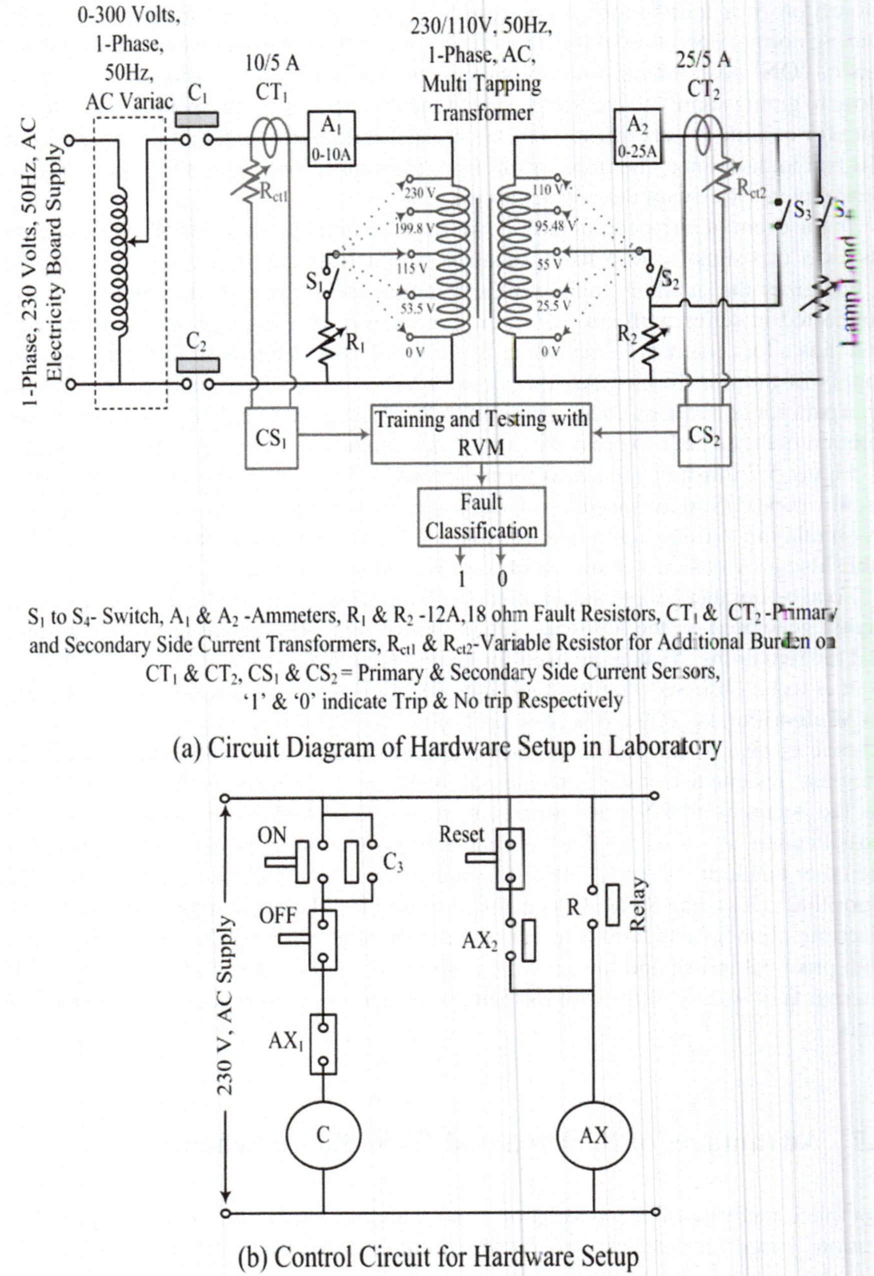

Fig. 5.7 Circuit diagram and control circuit of hardware setup

Table 5.9 Fault data generation using hardware setup

Fault cases	Inrush data	Training data (inrush)	Internal fault data	Training data (internal)	External fault data	Training data (external)
Inrush at different inception angle	6	4	–	–	–	–
Turn to turn	–	–	3	3	–	–
Fault location	–	–	3	2	3	2
Fault type	–	–	1 (L-G)	1	1 (L-G)	1
Fault resistance (R_f)	–	–	3	3	3	3
CT saturation during external fault	–	–	–	–	3	3
Total and training data	**6**	**4**	**27**	**18**	**27**	**18**
Testing data	**2**		**9**		**9**	

Bold represents a number of data/parameter used for the validation of the proposed scheme

1. The Bayesian formulation of the RVM circumvents the setting of the margin trade-off and the insensitivity parameter that requires in the SVM. Thus, cross-validation-based post-optimization is not required in RVM.
2. While implementing both classifiers for transformer fault discrimination, it is observed that the required support vectors (SVs) are much higher in SVM compare to relevance vectors (RVs) for RVM. This reduces computational complexity and also the time for classification is lesser in RVM compare to SVM and PNN.
3. As depicted in Table 5.6, the proposed RVM based scheme provides higher classification accuracy (more than 99%) during discrimination of all types of internal faults in the transformer. This shows the consistency of RVM techniques to identify and clear internal fault as early as possible.
4. Simultaneously, the proposed RVM based scheme offers almost 100% accuracy during all external faults and inrush condition as shown in Table 5.6. This feature avoids unnecessary outage of the transformer and remains stable during all external abnormalities.
5. It is to be noted from Table 5.7 that the fault classification accuracy obtained for different types of fault is also very high (more than 99%) particularly during internal winding fault and the external fault with CT saturation.

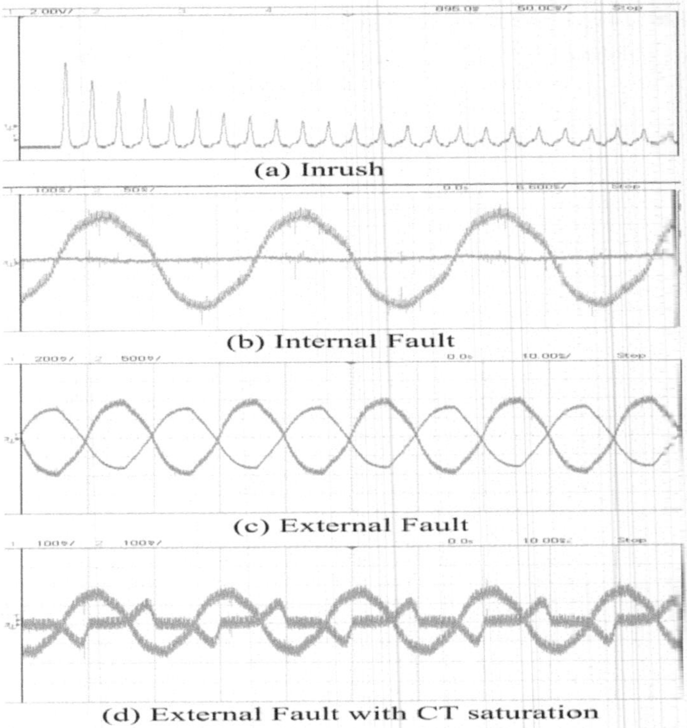

Fig. 5.8 Primary and secondary current waveform for **a** inrush **b** internal fault **c** external fault **d** external fault with CT saturation

6. It is observed during testing that the RVM technique represents higher accuracy with lesser time concerning SVM and PNN as shown in Table 5.8. This precision is obtained by considering 23,094 test cases.
7. The proposed RVM based technique is better in terms of classification accuracy and decision time than SVM and PNN based technique due to the Sparse Bayesian Learning theory. Furthermore, it does not require any pre or post-processing of the captured current signals and hence, its performance can be found better than method needs phasor and frequency estimation.
8. This technique is validated on hardware set up using RVM with SVM and PNN algorithm. It is found that the time of operation of the RVM scheme is lesser than SVM and PNN scheme due to fewer no's of RVs compare to SVs.

5.8 Summary

A new RVM based classifier scheme is proposed in this perspective to discriminate internal fault, external fault, and other abnormal conditions in power transformer

A part of the power system is simulated in PSCAD software to generate enormous data for validation of the proposed algorithm considering wide variation in system and fault parameters. One cycle post fault current samples of CT secondary are acquired from both sides of the transformer under consideration at 4 kHz sampling frequency. Out of a total of 53,784 cases, around 23,094 fault cases (42.94%) have been utilized for validation/testing of the proposed transformer protection technique. The proposed RVM based classifier scheme is compared with the existing SVM and PNN based classifier method. It is observed from the result that RVM provides higher efficiency and requires less time to classify faults and inrush current in transformer compare to SVM and PNN classifier. Moreover, the proposed scheme provides better reliability in classification of transformer internal fault by giving an accuracy of more than 99% and during inrush conditions and different types of external fault, fault classification accuracy is approximately 100%. It is to be mentioned here that to generate a strong trained model of RVM, past fault data are to be collected from the field or manufacturer. To check the feasibility of the proposed scheme, hardware-based fault data are generated in the laboratory. This scheme provides around 100% fault classification accuracy for the practical data and discriminates faults within the time prescribed.

5.9 Published Article Based on This Work

- Dharmesh D. Patel, N. G. Chothani, K. D. Mistry, and M. Raichura, "Design and Development of Fault Classification Algorithm based on Relevance Vector Machine for Power Transformer Design and Development of Fault Classification Algorithm based on Relevance Vector Machine for Power Transformer," *IET Electr. Power Appl.*, vol. 12, no. 4, pp. 557–565, 2018.

Appendix

Simulation data:

Line data-1	:	Length=100 km, system voltage = 400 kV Positive-sequence impedance = 0.0297 + j0.332 Ω/km Zero-sequence impedance = 0.162 + j1.24 Ω/km
Line data-2	:	Length = 80 km, System voltage = 220 kV Positive-sequence impedance = 0.032 + j0.456 Ω/km Zero-sequence impedance = 0.032 + j1.19 Ω/km
Transformer data	:	YY connected, 345 MVA, 400/220 kV, 3-phase, with 0.1 pu leakage reactance. (220 kV side of the transformer is connected to the infinite bus through line-2)

(continued)

(continued)

CT data	:	Primary-1000/5 A, Secondary-1800/5 A, Secondary winding resistance, and inductance = 0.5 Ω and $0.8e^{-3}$ H

Equipment data for hardware:

Transformer data	:	2 kVA, 230/115 V, 1-phase, 50-Hz, %Z = 12
CT data	:	Primary side: 10/5 A, 15 VA, 5p10 and for secondary side 20/5 A, 15 VA, 5p10
Load	:	Lamp load, 25 A
Source data	:	1-phase, 0–300 V, 50-Hz, variable supply from the electricity board

References

1. Tripathy M, Maheshwari RP, Verma HK (2007) Probabilistic neural-network-based protection of power transformer. IET Electr Power Appl 1(5):793–798
2. Balaga H, Gupta N, Vishwakarma DN (2015) GA trained parallel hidden layered ANN based differential protection of three phase power transformer. Int J Electr Power Energy Syst 67:286–297
3. Mittal M, Bhushan M, Patil S, Chaudhari S (2013) Optimal feature selection for SVM based fault diagnosis in power transformers. IFAC Proc 46(32):809–814
4. Bigdeli M, Vakilian M, Rahimpour E (2012) Transformer winding faults classification based on transfer function analysis by support vector machine. IET Electr Power Appl 6(5):268–276
5. Koley E, Shukla SK, Ghosh S, Mohanta DK (2017) Protection scheme for power transmission lines based on SVM and ANN considering the presence of non-linear loads. IET Gener Transm Distrib 11(9):2333–2341
6. Ashrafian A, Rostami M, Gharehpetian GB (2012) Hyperbolic S-transform-based method for classification of external faults, incipient faults, inrush currents and internal faults in power transformers. IET Gener Transm Distrib 6(10):940–950
7. Gil M, Abdoos AA (2017) Intelligent busbar protection scheme based on combination of support vector machine and S-transform. IET Gener Transm Distrib 11(8):2056–2064
8. Saleh SA, Aktaibi A, Ahshan R, Rahman MA (2012) The development of a d-q axis WPT-based digital protection for power transformers. IEEE Trans Power Deliv 27(4):1–8
9. Chen J, Phung T, Blackburn T, Ambikairajah E, Zhang D (2016) Detection of high impedance faults using current transformers for sensing and identification based on features extracted using wavelet transform. IET Gener Transm Distrib 10(12):2990–2998
10. Medeiros RP, Costa FB, Silva KM (2016) Power transformer differential protection using the boundary discrete wavelet transform. IEEE Trans Power Deliv 31(5):2088–2095
11. Maya P, Vidya Shree S, Rupashree K, Soman KP (2015) Discrimination of internal fault current and inrush current in a power transformer using empirical wavelet transform. Procedia Technol 21:514–519
12. Ray P, Mishra DP (2016) Support vector machine based fault classification and location of a long transmission line. Eng Sci Technol Int J 19(3):1368–1380
13. Shah AM, Bhalja BR (2013) Discrimination between internal faults and other disturbances in transformer using the support vector machine-based protection scheme. IEEE Trans Power Deliv 28(3):1508–1515
14. Zhang LL, Wu QH, Ji TY, Zhang AQ (2017) Identification of inrush currents in power transformers based on higher-order statistics. Electr Power Syst Res 146(Supplement C):151–169

15. Tipping ME (2001) Sparse Bayesian learning and the relevance vector machine. J Mach Learn Res 1:211–244
16. Naveen N, Natarajan S, Srinivasan R (2012) Application of relevance vector machines in real time intrusion detection. Int J Adv Comput Sci Appl IJACSA 3(9):48–53
17. Li R (2012) Computer network attack evaluation based on incremental relevance vector machine algorithm. J Convergence Inf Technol JCIT 7(1):43–48
18. Lou J, Jiang Y, Shen Q, Shen Z, Wang Z, Wang R (2016) Software reliability prediction via relevance vector regression. Neurocomputing 186:66–73
19. Niu L, Zhao J-G, Li K-J (2012) Application of data mining technology based on RVM for power transformer fault diagnosis. Adv Intell Soft Comput 169:121–127
20. Yin J, Zhou X, Ma Y, Wu Y, Xu X (2015) Power transformer fault diagnosis based on multi-class multi-kernel learning relevance vector machine. In: 2015 IEEE international conference on mechatronics and automation (ICMA), pp 217–221
21. PSCAD Research Center (2005) EMTDC-transient analysis for PSCAD power system simulation. Winnipeg, MB, Canada
22. van Warrington AR (1962) Protective relays–their theory and practice, vol 1, 1st edn. Chapman and Hall, London
23. van Warrington ARC (1968) Protective relays: their theory and practice/by A. R. van C. Warrington, 2nd edn. Chapman & Hall London
24. Mohanty SR, Pradhan AK, Routray A (2008) A cumulative sum-based fault detector for power system relaying application. IEEE Trans Power Deliv 23(1):79–86
25. Mohebian R, Riahi MA, Afjeh M (2018) Detection of the gas-bearing zone in a carbonate reservoir using multi-class relevance vector machines (RVM): comparison of its performance with SVM and PNN. Carbonates Evaporites 33(3):347–357

Chapter 6
HE-ELM Technique Based Transformer Protection

Various unwanted phenomena that are taken place in the transformer may occasionally mal-operate selected fault classification based protective schemes. Hence, it is necessary to discriminate internal fault from external abnormal conditions for unit protection of power transformer. This paper presents a new Hierarchical Ensemble Extreme Learning Machine (HE-ELM) based classifier technique to identify faults in and out of the transformer. The component ELMs is structured hierarchically to improve its fault data classification accuracy. The developed algorithm is evaluated by simulating multiple disorders on 100 MVA, 132/220 kV transformer with the help of PSCAD software. DWT is used to extract features from acquired current signals from the transformer. The feature vector formed after the extraction process is fed to the HE-ELM algorithm for data classification. The fault discrimination accuracy of the HE-ELM technique is 99.91%. This shows its effectiveness concerning other classifier techniques. Moreover, the developed algorithm is successfully tested on hardware prototype in a laboratory environment under various inrush and fault conditions using a Cortex M4 microcontroller (STM32F407) with maximum identification time of 27 ms. The proposed HE-ELM technique is compared with existing SVM, PNN, and ELM techniques for identical fault data. Results demonstrate that HE-ELM outperforms than existing schemes in the cross-domain recognition task.

6.1 Documentation of Comprehensive Review

Protection of power transformer is one of the most complicated tasks in the power system protection field, because of the nonlinear magnetic characteristics of the core and its construction. Moreover, complexity increases with different voltage/current ratios for unit type protection. Some techniques present good fault classification accuracy among decomposing and filtering methods. Specific parameterizations affect on the problem statement, classifier design, and its performance. Many researchers have effectively proposed various classifier techniques for transformer protection with an

D. Patel and N. Chothani, *Digital Protective Schemes for Power Transformer*, Power Systems, https://doi.org/10.1007/978-981-15-6763-6_6

optimized parameter like SVM, LSSVM, PNN, GA, and ANN. Though, the research area in the field of transformer protection has intensified some lacking points to raise research gaps.

Support Vector Machine (SVM) based fault classification technique for power transformer protection has been proposed by Shah et al. [1]. However, this scheme takes more calculation time in case of large training data set, and also the classification accuracy varies with different transformer connections. Extreme Learning Machine (ELM) theory and it's applications have been first proposed by Huang et al. [2]. Further, Jan Wang et al. [3] and Dogaru et al. [4] presented a comparison of SVM and ELM based classifier technique, in which they have proved that ELM outperforms than SVM. Hence, from the mentioned references one can say that ELM is better than SVM in the application point of view as well as accuracy concern. Also, Koroglu et al. [5] presented a diagnosis of power transformer fault depending on the oil deterioration and dissolved gas analysis. They found the severity of fault and damage that occurred to the insulation. However, the scheme provides only 92% diagnostic accuracy, and also they didn't perform discrimination of inrush and fault condition.

A variety of other classification-based techniques were proposed by various researchers to classify the internal fault and other abnormalities correctly, but all those methods either require high computational time or provide less classification accuracy. Probabilistic Neural Network (PNN) classifier with the help of Principle Component Analysis (PCA) has been proposed in [6], which also concern with the percentage accuracy for fault and abnormalities discrimination. Genetic Algorithm (GA) trained to parallel hidden layered Artificial Neural Network (ANN) based differential protection of a three-phase power transformer has been proposed by Balanga et al. in [7]. This scheme requires more computational time, as the proposed algorithm has to follow 7 different steps every time during its training period. Relevance Vector Machine (RVM) based classification technique has been implemented in the power transformer protection purpose by Chothani et al. [8]. The method gives higher accuracy than SVM. However, this classifier method may suffer from high computational time in the case of bulky training data set.

Besides classification-based techniques, many researchers suggested various other techniques that can discriminate between transformer internal fault and external abnormalities. Maya et al. [9] presented transformer protection using Empirical Wavelet Transform, however, they did not discuss all the abnormal conditions which are taken place during the transformer operation. Moreover, Chen J. et al. presented the detection of High Impedance Fault (HIF) with the help of WT in [10]. However, accuracy noted is only 72% even by considering 3 full cycle waveforms. A superimposed differential current based protection scheme has been proposed by Shah et al. in [11]. In this article, the authors used a time-time transform to detect a fault condition. With the help of superimposed current, they have identified external or internal abnormalities. However, the method is not capable to detect transformer internal LLL fault and LLL-G fault.

Internal fault fast identification criterion based on superimposed component comparison for power transformer has been presented by Lin et al. [12]. They have utilized time interval between voltage and current to detect the fault and other disturbances. However, this scheme is valid for the sudden change in voltage and not capable to detect HIF and incipient fault conditions. Bridge type fault current limiter (BFCL) has been used in [13] to retain the sensitivity of restricted earth fault protection scheme. Though, during fault near the transformer neutral, the proposed method cannot protect the asset. Dashti et al. [14] elaborated discrimination of large inrush current from fault currents with the help of assessment of symmetry between two zeros of a cycle and by introducing a different function but, the method alone is not capable to detect low or mild inrush currents. Transformer differential protection with considering Current Transformer (CT) saturation and cross country fault has been proposed by Medeiros and Costa [15]. The entire concept is based on the wavelet energy of the differential current. The proposed method detects the energization event and at the same time if the fault is already present, then it can't detect that fault, because the logic identifies it as the energization event. This limitation addressed in the article [16] by the same authors. In [16], the discrimination margin is very less between operating WT energy and restraining WT energy which may mal-operate unforeseen conditions like heavy CT saturation or high resistance fault conditions.

Setting-free differential protection for power transformers based on the Second Central Moment (SCM) has been proposed by Esponda et al. [17]. The method calculates the SCM using the integration of a waveform. During the inrush condition, the magnitude of SCM remains below threshold (0.25), while during fault condition the value of SCM will be greater than the threshold. But during heavy CT saturation condition, though the waveform is bidirectional it is distorted and having a lower magnitude than the threshold value, which may mal-operate the relay.

Magnetization hysteresis based power transformer protection has been proposed by Zaibin Jiao et al. [18]. The authors observed that the B-H curve decline during a fault condition and this inclination can be identified by the SVM classifier method. To obtain the B-H curve both CTs and Potential Transformers (PTs) are used which may increase the cost and dependency of the protection scheme. Nima Farzin et al. proposed transformer Turn-to-Turn Fault (TTF) detection based on Fault-Related Incremental Currents (FRIC) [19]. If the difference of sequence component before and after fault increases the preset threshold, then it can be interpreted as the existence of TTF. The method fails to detect the TTF fault condition that exists before transformer energization, as the FRIC scheme is bypassed during inrush conditions.

A new algorithm based on the Hierarchical Ensemble Extreme Learning Machine (HE-ELM) [20] classifier technique to discriminate power transformer internal fault and external fault or abnormal conditions are presented here. HE-ELM is an improved version of ELM. HE-ELM improves the diversity of component ELMs which reduces the overfitting of ELM at the time of training. The feature bagging method used in HE-ELM also reduces the computational complexity of ELM, which is described in the subsequent section. The proposed technique provides acceptable results and can be utilized as a modern power transformer protection scheme. Power System Computer-Aided Design (PSCAD^{TM}) [21] software is utilized for system modelling and data collection with variation in parameters of the power system. All data generated in

PSCAD™ software is tested and verified with the HE-ELM algorithm developed in MATLAB software. Section 6.2 describes system modelling and generation of total data. Sections 6.3 and 6.4 details the proposed technique and learning model developed. Section 6.5 shows the proposed fault classification algorithm. Sections 6.6, 6.7 depicts validation of the proposed scheme on PSCAD software, comparison with the existing schemes. Sections 6.8, 6.9 and 6.10 elaborated hardware setup, its test results additional hardware results and benefits of the proposed scheme sequentially.

6.2 System Modeling, Data Generation and Simulation

As shown in Fig. 6.1, a part of the Indian power system of Karamsad substation having three power transformers of 132/220 kV, 100MVA (Thevenin's equivalent) are considered for the simulation. Here, a 132 kV, 80 km transmission line from Dhuvaran generating station is connected to the primary of the paralleled transformer, and a 220 kV, 30 km transmission line from Kasor substation is connected on the secondary side. CT_{11}, CT_{12}, CT_{13} and CT_{21}, CT_{22}, CT_{23} represent a set of Current Transformers (CTs) on primary as well as secondary of each power transformer, respectively. 115% overloading of each transformer is considered for fixing the rating of CTs. The said existing network of the power system is simulated in PSCAD™ using actual parameters of line and transformers as collected from the field. A large number of fault cases and abnormalities are created to test the developed HE-ELM, SVM, PNN, and ELM based classifier techniques. A wide variation in system parameters is chosen such as type of fault (F_{type}), different Source Impedances (SI), varying Fault Inception Angle (FIA), Fault resistance (R_f), load angle (δ) as well as Fault Locations (*FLs*) on transformer winding (F_{int}) and transmission line/bus (F_{ext}). To

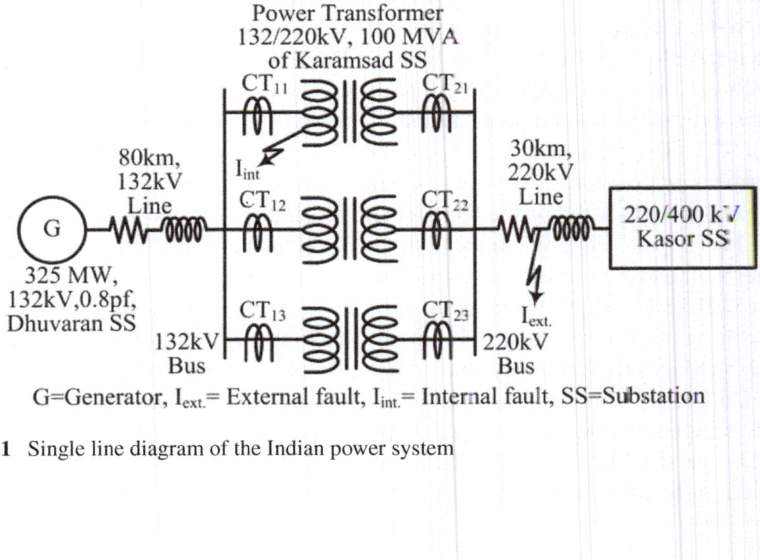

Fig. 6.1 Single line diagram of the Indian power system

produce numerous simulation cases, a multi-run block available in PSCAD [21] is utilized to simulate a large number of cases simultaneously.

Variation in different parameter values is shown in Table 6.1 for internal fault conditions, Table 6.2 for external fault conditions, and Table 6.3 for various inrush conditions with the division of training and testing data. Table 6.1 shows a total of 48,720 cases for transformer internal fault with various parameters including TTF, inter winding, and internal faults. Out of total data generated, 29 232 data (60%) are considered as training data and 19,488 data (40%) are taken as testing data. Similarly, Table 6.2 shows a total of 30,000 cases for an external fault occurring outside the transformer zone, among them 21,000 data (70%) are considered as training data, and the remaining 9000 data (30%) are used as testing data. Inrush conditions have prime

Table 6.1 Training and testing data generated through various internal fault conditions

Parameter variation	Turn to turn fault	Turn to turn fault (training data)	Primary to secondary winding fault	Primary to secondary winding fault (training)	Internal winding fault	Internal winding fault (training)
Fault type (F_{type})	6 (3-on primary, 3-on secondary)	6 (3-on primary, 3-on secondary)	3 (all in three winding)	3 (all in three winding)	10 (L-g, LL, LL-g, LLL)	10 (L-g, LL, LL-g, LLL)
Source impedance (SI)	3 (80%, 100%, 120%)	3 (80%, 100%, 120%)	3 (80%, 100%, 120%)	3 (80%, 100%, 120%)	3 (80%, 100%, 120%)	3 (80%, 100%, 120%)
Fault location (FL)	7 (0.2%, 1%, 2%, 3%, 4%, 5%, 7%)	6 (0.2%, 1%, 2%, 3%, 4%, 7%)	7 (0%, 15%, 30%, 45%, 60%, 75%, 90%)	6 (0%, 15%, 30%, 60%, 75%, 90%)	7 (0%, 15%, 30%, 45%, 60%, 75%, 90%)	6 (0%, 30%, 45%, 60%, 75%, 90%)
Fault inception angle (FIA)	8 (0°, 15°, 30°, 60°, 90°, 120°, 135°, 150°)	7 (0°, 15°, 30°, 60°, 120°, 135°, 150°)	8 (0°, 15°, 30°, 60°, 90°, 120°, 135°, 150°)	7 (0°, 15°, 30°, 60°, 120°, 135°, 150°)	8 (0°, 15°, 30°, 60°, 90°, 120° 135°, 150°)	7 (0°, 15°, 60°, 90°, 120°, 135°, 150°)
Fault resistance (R_f)	–	–	4 (0 Ω, 5 Ω, 10 Ω, 15 Ω)	4 (0 Ω, 5 Ω, 10 Ω, 15 Ω)	4 (0 Ω, 5 Ω, 10 Ω, 15 Ω)	4 (0 Ω, 5 Ω, 10 Ω, 15 Ω)
Load angle (δ)	5 (0°, 5°, 10°, 15°, 20°)	4 (0°, 5°, 15°, 20°)	5 (0°, 5°, 10°, 15°, 20°)	4 (0°, 5°, 15°, 20°)	5 (0°, 5°, 10°, 15°, 20°)	4 (0°, 5°, 15°, 20°)
Total	5040	Training data = 3024, testing data = 2016	10,080	Training data = 6048, testing data = 4032	33,600	Training data = 20,160, testing data = 13,440

Hence, from total 48,720 data for transformer internal fault, training data = 29,232 and testing data = 19,488

Table 6.2 Training and testing data for various external faults

Parameter variation	Fault on 220 kV bus (with and without CT saturation)	Fault on 220 kV bus (with and without CT saturation) (training Data)	Fault on 220 kV line	Fault on 220 kV line (training data)
Fault type (F_{type})	10 (L-g, LL-g, LL, LLL)	10 (L-g, LL-g, LL, LLL)	10 (L-g, LL-g, LL, LLL)	10 (L-g, LL-g, LL, LLL)
Source impedance (SI)	3 (80%, 100%, 120%)	3 (80%, 100%, 120%)	3 (80%, 100%, 120%)	3 (80%, 100%, 120%)
Fault location (FL)	–	–	3 (1 km, 20 km, 30 km)	3 (1 km, 20 km, 30 km)
Fault inception angle (FIA)	8 (0°, 15°, 30°, 60°, 90°, 120°, 135°, 150°)	7 (0°, 15°, 60°, 90°, 120°, 135°, 150°)	8 (0°, 15°, 30°, 60°, 90°, 120°, 135°, 150°)	7 (0°, 15°, 60°, 90°, 120°, 135°, 150°)
Fault resistance (R_f)	5 (0 Ω, 5 Ω, 10 Ω, 15 Ω, 20 Ω)	5 (0 Ω, 5 Ω, 10 Ω, 15 Ω, 20 Ω)	5 (0 Ω, 5 Ω, 10 Ω, 15 Ω, 20 Ω)	5 (0 Ω, 5 Ω, 10 Ω, 15 Ω, 20 Ω)
Load angle (δ)	5 (0°, 5°, 10°, 15°, 20°)	4 (0°, 5°, 10°, 20°)	5 (0°, 5°, 10°, 15°,20°)	4 (0°, 5°, 10°, 20°)
Total	6000 * 2 = 12,000	Training data = 4200 * 2 = 8400, testing data = 3600	18,000	Training data = 12,600, testing data = 5400

Hence, from a total of 30,000 data for external fault outside the transformer zone, training data = 21,000 and testing data = 9000

importance in unit type protection of the transformer due to mal-operation of existing schemes. Table 6.3 shows the separation of training and testing data by considering various inrush conditions such as initial inrush, sympathetic inrush, and recovery inrush. Total 1200 inrush cases are taken into account as various magnetizing inrush conditions in transformer, among them 702 data (58.5%) are training data and 498 data (41.5%) are testing data.

It is proved that for getting better accuracy in any classifier, training data must be larger compare to testing data [4]. Also, the selection of proper training data and testing data is most important in some techniques. Among 79,920 total data, 50,934 data (63.73%) are considered as training data and 28,986 data (36.27%) are considered as testing data. Table 6.4 shows an overview of total data generated by simulating internal fault conditions, external fault, external fault with CT saturation conditions, and various inrush conditions.

Table 6.3 Training and testing data generated for various inrush conditions

Parameter variation	Initial inrush	Initial inrush (training data)	Sympathetic inrush	Sympathetic inrush (training data)	Recovery inrush	Recovery inrush (training data)
Source impedance (SI)	3 (80%, 100%, 120%)	3 (80%, 100%, 120%)	3 (80%, 100%, 120%)	3 (80%, 100%, 120%)	3 (80%, 100%, 120%)	3 (80%, 100%, 120%)
CB switching instant	10 (0°, 15°, 30°, 45°, 60°, 75°, 90°, 105°, 120°, 150°)	9 (0°, 15°, 30°, 45°, 60°, 90°, 105°, 120°, 150°)	10 (0°, 15°, 30°, 45°, 60°, 75°, 90°, 105°, 120°, 150°)	9 (0°, 15°, 30°, 45°, 60°, 90°, 105°, 120°, 150°)	10 (0°, 15°, 30°, 45°, 60°, 75°, 90°, 105°, 120°, 150°)	9 (0°, 15°, 30°, 45°, 60°, 90°, 105°, 120°, 150°)
Load angle (δ)	5 (0°, 5°, 10°, 15°, 20°)	4 (0°, 5°, 15°, 20°)	5 (0°, 5°, 10°, 15°, 20°)	3 (0°, 15°, 20°)	5 (0°, 5°, 10°, 15°, 20°)	3 (0°, 15°, 20°)
Residual flux	6 (0%, 10%, 25%, 45%, 60%, 80%)	5 (0%, 10%, 45%, 60%, 80%)	–	–	–	–
Total	900	Training data = 540, testing data = 360	150	Training data = 81, testing data = 69	150	Training data = 81, Testing data = 69

Hence, from total 1200 data for various magnetizing inrush condition in transformer, training data = 702 and testing data = 498

Table 6.4 Total training and testing data collection for various conditions

Cases	Training data	Testing data	Total data
Internal fault	29,232	19,488	48,720
External fault with and without CT saturation	21,000	9000	30,000
Inrush	702	498	1200
Total	**50,934**	**28,986**	**79,920**

Bold represents a number of data/parameter used for the validation of the proposed scheme

6.3 Existing and Proposed Techniques for Transformer Protection

Binary classification is formulated to discriminate the category of faults such as internal, external, inrush, and normal conditions. The fault data generated in PSCAD are migrated to MATLAB to form feature vectors after feature extraction for one cycle post-disturbance duration. The accuracy of the HE-ELM classifier is compared with other techniques such as ELM, PNN and SVM using m-code in MATLAB software.

6.3.1 PNN Learning Model

PNN is a pattern recognition classifier technique of feed-forward NN. Here Gaussian based function is utilized as activation function:

$$f_{ij}(x; c_{ij}, \sigma) = \frac{1}{(2\pi)^{d/2}\sigma^{d}} exp\left(-\frac{1}{2\sigma^2}(x - c_{ij})^T(x - c_{ij})\right) \quad (6.1)$$

where, x is the test input vector, T denotes the transpose of the vector, $i = 1, 2, \ldots, n$ and $j = 1, 2, \ldots, M$ (M = number of pattern unit) also here standard deviation (σ) is defined as smoothing factor, c is centers of the Kernel and d is dimensionality.

Here $M = \sum_{i=1}^{k} M_i$ and $\sum_{j=1}^{M_i} w_{ij}$ for every given class, k_i where $i = 1, 2, \ldots n$. Means each layer summation at each node estimates the conditional probability destination function $p_i(x/k_i)$ of each class of k_i, defined as:

$$p_i(x/k_i) = \frac{1}{(2\pi)^{d/2}\sigma^{d}} \frac{1}{M_i} \sum_{j=1}^{M_i} \exp\left(-\frac{1}{2\sigma^2}(x - x_{ij})^T(x - x_{ij})\right) \quad (6.2)$$

where x_{ij} is, jth training vector for class k_i, d is the dimension of the feature vectors and M_i is the number of training pattern in class k_i. So the output layer of PNN is known as a competitive layer.

The structural parameters of PNN

The classification accuracy of the PNN depends on the value of the smoothing factor (σ). If the value of σ is too large or small, the network will converge too fast or fail into locally optimal solution. The conventional trial-and-error method is used to obtain a smoothing factor. Here, we get 0.53 as an optimum value of the smoothing factor after the trial-and-error method. Though there are certain algorithms available to derive the optimum value of smoothing factor [22–24], instead of manual exercise. In that case, also some parameters (like multiplying parameter (g)) have to be found by experiments [22], hence we choose to manually select the value of the smoothing factor. Other structural parameters are chosen based on the numbers of input cases (dimension of feature space), a number of classification types, and number of decisions to be obtained. The typical PNN structure used in this work is shown in Fig. 6.2. It is a four-layer feed-forward neural network that is capable of realizing or approximating the optimal classifier.

Further, many researchers have introduced ANN and PNN in the power system for fault data classification directly with the sampled voltage and current signals

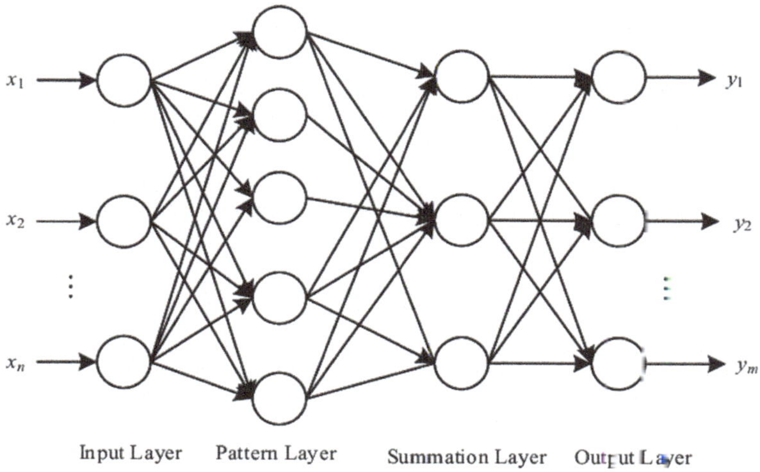

Fig. 6.2 Structure of PNN

[22, 23, 25]. Moreover, SVM can be trained without any pre-processing or feature extraction as it utilized sampled data of current signals for classification. To compare the suggested technique (HE-ELM) and past techniques (SVM, PNN) for transformer fault classification, authors have considered uniform methodology (classification without feature extraction) for training for all the methods discussed in this work [26–28].

Moreover, as far as a matter of comparison of simple feed-forward ANN and PNN is concerned, there are some kinds of literature available in the same field of transformer protection. Researchers have already shown that the performance of PNN is better than simple feed forwarded ANN for fault classification [22, 24]. So, we can definitely say that the performance of the PNN for this work is obviously better than a simple feed forwarded ANN technique based classification. Also, the space limitation of the article, as per the journal guidelines restricts the inclusion of all the detail about simple feed forwarded ANN. Hence, we have not included the comparison of PNN with simple feed forwarded ANN but we have mentioned the adequate reference here, which can prove that PNN performs better in terms of classification accuracy as well as in testing time than simple feed forwarded ANN.

6.3.2 SVM Learning Model

Based on Optimal Separating Hyper-Planes (OSHP), SVM performs the classification of testing data. OSHP belongs to two separate classes based on a maximum margin between two data points. For this event inequality valid for all input data:

$$y_i\left(w^T x_i + b\right) \geq 1, \quad \text{For all } x_i, i = 1, 2, \ldots n. \tag{6.3}$$

Optimal bias is given by

$$b^* = y_i - w^{*T} x_i \tag{6.4}$$

Here $x_i =$ support-vector and optimal decision function is given as

$$f(x) = sgn\left(\sum_{i=1}^{n} \left(y_i \alpha_i x^T x_i + b^*\right)\right) \tag{6.5}$$

where α_i are optimal Lagrange multiplier and SVM used with soft margin along with non-negative slack variables (ζ_i) for high noise input given by:

$$\zeta_i = 1, 2, \ldots, n$$

$$y_i\left(w^T x_i + b\right) > 1 - \zeta_i \quad \text{For } i = 1, 2 \ldots n \tag{6.6}$$

For obtaining OSHP, it should decrease the

$$\emptyset = \frac{1}{2} w^2 + C \sum_{i=1}^{l} \zeta_i^k \tag{6.7}$$

where C is the penalty parameter to control complexity between decision function and training examples to avoid misclassification.

For nonlinear cases, SVM maps training points to a higher dimension feature using a Kernel function

$$K\left(x_i, x_j\right) = exp\left(\frac{\|-x_i - x_j\|}{2\sigma^2}\right) \tag{6.8}$$

where σ is a parameter of the kernel function.

After training, decision function is defined as,

$$f(x) = sgn\left(\sum_{i=1}^{l} y_i \alpha_i^* K(x, x_i) + b^*\right) \tag{6.9}$$

SVM performance is controlled with the terms C and kernel parameter called hyperparameters. SVs and margin maximization make an influence on the decision of SVM.

6.3.3 ELM Learning Model

ELM is mathematically represented, when L hidden layer neurons for sample data $\{x_i, y_i\}_{i=1}^{N}$ is

$$\sum_{i=1}^{L} \beta_i G(a_i, b_i, x_i) = y_i, \quad i = 1, 2, \ldots, N \tag{6.10}$$

where a_i and β_i are a vector of input and output weight respectively, G is activation function and b_i denotes bias of ith hidden node.

For convenience, we can rewrite as

$$\|H\beta = Y\| \tag{6.11}$$

where $H = \begin{bmatrix} h_1(x_1) & \ldots & h_L(x_1) \\ \vdots & \vdots & \vdots \\ h_1(x_N) & \ldots & h_L(x_N) \end{bmatrix}$

Here, H is the hidden layer output matrix.

For minimizing training error and norms of the output weight in ELM [2].

Minimize $\|H\beta - Y_2\|$ *and* $\|\beta\|$.

$\beta = \begin{bmatrix} \beta_1^T \\ \vdots \\ \beta_N^T \end{bmatrix}$ And $Y = \begin{bmatrix} y_1^T \\ \vdots \\ y_N^T \end{bmatrix}$

Traditionally, to train SLFN specific weight factor (w_i), the threshold of the ith hidden node (b_i), weight factor (β) are connected to an ith hidden node in such a manner that,

$$\left\| H\left(w_{1,\ldots,}w_N, b_{1,\ldots,}b_N\right)\beta - Y \right\| = \min_{w_{i,b_i,\beta}} \left\| H\left(w_{1,\ldots,}w_N, b_{1,\ldots,}b_N\right)\beta - Y \right\| \tag{6.12}$$

This is equal to minimizing cost factor to improve accuracy,

$$E = \sum_{j=1}^{N}\left(\sum_{i=1}^{N} \beta_i g(w_i x_i + b_i) - y_j\right)^2 \tag{6.13}$$

where E is unknown gradient-based learning algorithms, search the minimum of $H\beta = Y$.

To minimize the gradient-based algorithm, weight factor and other parameter are adjust as follow:

$$W_k = W_{k-1} - \gamma \frac{dE(W)}{dW} \tag{6.14}$$

Here γ = learning rate and vector W is the set of the weight $(wi, \beta i)$. Here accuracy is computed by propagation from output to input.

6.4 Proposed HE-ELM Learning Model

Hierarchical Ensemble Extreme Learning Machine (HE-ELM) is an updated version of ELM, which provides enhanced classification accuracy. Hence, with the help of the hierarchical structure ensemble of ELM has been built [20].

HE-ELM structure consists of 2 re-representation layers (composed of ELM) and a decision layer. After training of the first re-representation layer output of it is generated through the feature bagging method.

In the first re-representation layer, i_a component ELM will be trained separately using the same features which are used for training of ELM, as discussed in the previous subsection. These trained components of ELMs are, denoted as $A_1 = \{a_i^{(1)}\}_{i=1}^{i_a}$. For ith component $ELM\ a_i^{(1)}$, hidden layer components are randomly initialized which is represented as $\varphi_i^{(1)}$. The prediction vector of x_k predicted by ith component ELM as follows,

$$p_i^{(1)} = a_i^{(1)}(x_k; \varphi_i^{(1)}) \tag{6.15}$$

Here, $p_i^{(1)}$ is a C-dimensional vector of continuous values. By considering all predictions of x_k derived by A_1, and associate them with the input of the first re-representation layer is trained. Here, re-representation of x_k is denoted as

$$\bar{x}_k^{(1)} = [x_k; p_1^{(1)}, p_2^{(1)}, \ldots, p_{i_a}^{(1)}] \tag{6.16}$$

$\bar{x}_k^{(1)}$ is a $(i + C \cdot i_a)$- dimensional vector.

The second re-representation vector can be similarly given as $A_2 = \{a_i^{(2)}\}_{i=1}^{i_b}$ which tends to train the latest re-representation for $\bar{x}_k^{(1)}$ using i_b component ELMs. The dimension $\bar{x}_k^{(1)}$ is most probably higher compared to initial input dimensions, which results in redundancy and increases computational complexity, and hence feature bagging is used to sample from sub-space before the second re-representation layer. The Feature bagging is a method in which training of component ELMs has been done directly on subset features of input rather than the training of whole feature space. This will reduce the risk of over fitting of HE-ELM. Moreover, the reduced features make the component ELMs more compact which helps in reduced training time. This is given as, the feature subspace sampling is repeated independently with equal probability for i_s times, and k feature is randomly selected from $\bar{x}_k^{(1)}$ for each time, where k is given as,

$$k = [0.6 \cdot (i + C \cdot i_a)] \tag{6.17}$$

Here, one thing is good to note that the feature bagging procedure generates i_s sub-samples from which only 60% of information is kept. These sub-samples are then passed to A_2 and produce a prediction matrix, which is expressed as,

$$O^{(2)} = \begin{bmatrix} o_{11}^{(2)} & o_{12}^{(2)} & \cdots & o_{1i_s}^{(2)} \\ o_{21}^{(2)} & o_{22}^{(2)} & \cdots & o_{2i_s}^{(2)} \\ \cdots & \cdots & \cdots & \cdots \\ o_{i_b1}^{(2)} & o_{i_b2}^{(2)} & \cdots & o_{i_bi_s}^{(2)} \end{bmatrix} \tag{6.18}$$

Here, $O^{(2)} \in \mathrm{R}^{i_b \times i_s \times C}$, in which ith row shows the prediction of i_s sub-samples predicted by $a_i^{(2)}$ which can be represented as

$$P_i^{(2)} = a_i^{(2)}(\bar{x}_k^{(1)}; \varphi_i^{(2)}) \tag{6.19}$$

After that, we vectorize $O^{(2)}$ it is correlated with x_k. The new representation can be written as

$$\bar{x}_k^{(2)} = [x_k; vec(O^{(2)})] \tag{6.20}$$

where $vec\,()$ indicates vectorizing a matrix with rows.

To determine the final stage decision layer has been introduced. The output layer named, Ridge Regression classifier is taken as a decision layer which keeps the structure consistent. The final class can be derived with the help of the following process,

$$class(x_k) = \underset{c=1,\cdots,C}{argmax_c}(\bar{x}_k^{(2)}\beta) \tag{6.21}$$

Here, β is a weight matrix of the Ridge Regression classifier. This may be calculated from Eq. (6.11).

In the proposed work, the sparse connection is utilized to compact component ELMs. A simple ELM generally uses a big number of neurons to give an accurate performance. Based on [29] we have utilized Bernouli's probability S to randomly choose hidden connection between neurons, where S is called Sparsity rate. Significance of Sparsity rate S can be given as the ratio of disconnected connections between the input layer and hidden neurons of ELM.

6.4.1 Feature Extraction Using Wavelet Transform

Wavelet Transform (WT) is a sound feature extraction technique that reduces the dimensionality of the entire input data and kept the relative information as a feature [30]. Different mother wavelets are available for feature extraction like Harr,

Symmlet, Couflet, Daubichies etc. Feature extraction has been performed by the Discrete Wavelet Transform (DWT) method as this work dealt with discrete current signal waveforms. DWT method segregates the given data into details and approximation. The selection of mother wavelet is important to properly extract the feature. The Daubechies (*db*) wavelet is generally used to observe the fault transients since the characteristics shape of *db* is similar in the shape of fault transients [1]. In DWT, the time-scale representation of a discrete signal is obtained by a digital filtering technique [31]. Here, the first-level decomposition has been utilized with the help of the db4 mother wavelet.

6.5 Proposed Fault Classification Algorithm

Various disturbances are simulated on the considered Indian power system. The abnormality detection algorithm [32] discriminates between abnormal condition and the normal operating state. During the simulation, one cycle post-disturbance current samples are separated to form a feature vector. The data sampling is done at 4 kHz sampling frequency (80 samples/cycle) [33]. The sampled data is given to DWT for feature extraction. The training data set as described in Tables 6.1, 6.2, 6.3 and 6.4 is used for offline training of the proposed HE-ELM algorithm. To train the classifier technique with Probabilistic Bayesian Learning (PBL), 50,934 fault cases (63.73% of total 79,920 cases) have been considered.

The algorithm for training and testing of the proposed classifier is depicted in Fig. 6.3. A feature vector of one post-disturbance cycle after feature extraction (DWT) is used for testing of HE-ELM classifier. The output of the classifier is divided into two categories as an internal fault (+1) or external abnormalities/normal conditions (−1). It is to be noted that parameter selection is a major task in all classifier techniques. Moreover, *K*-fold cross-validation is performed with the available fault data to check the authenticity of the proposed HE-ELM technique on unseen data (Table 6.8).

6.5.1 Parameter Tuning

The trained model is further utilized to validate the feature vector of test data.

Here, parameters of HE-ELM are optimized to get better accuracy of fault classification. In the proposed algorithm $ia = ib = ih = 20$, $is = 10$, S (sparsity rate) $= 0.2$ gives highest fault classification accuracy of 99.91% compare to the SVM technique with RBF kernel having parameters such as, gamma $g = 0.0415$ and regularization C $= 1000$, gives highest classification accuracy of 99.77%. Moreover, the smoothing factor (σ) of PNN has been taken as 0.53 optimally based on the trial-and-error method.

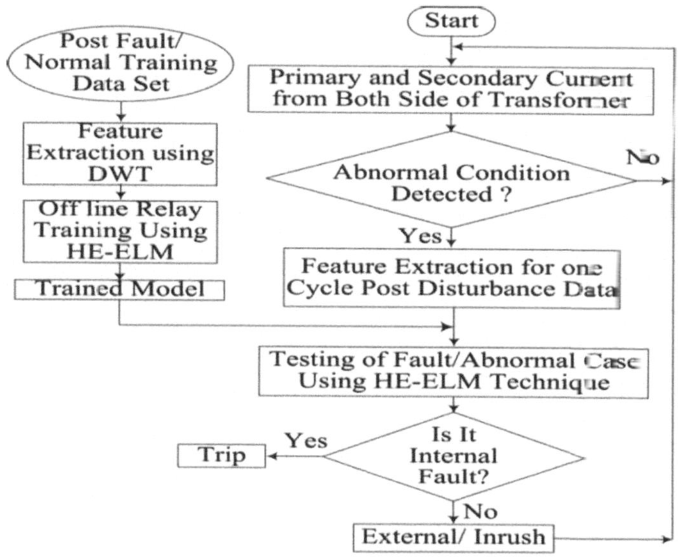

Fig. 6.3 Proposed HE-ELM technique based algorithm

6.6 Result Analysis and Discussion

6.6.1 Justification for Selection of the Size of Training Data Set in the Proposed Scheme

As mentioned in Table 6.4, the total data generated is 79,920. Now as per [4], the training data should be greater than testing data. So, to find the optimum number of training and corresponding testing data, authors have done assorted exercises by taking varying numbers of training and testing data. From Table 6.5, it can be seen that the optimum size of training data can be taken as 63.73% of the total (79,920) data. Below which will give lower accuracy and above which no significant improvement is found inaccuracy.

A very good thing to note, we can take a higher portion of training data (≥90% of total data) but the machine learning will take much higher time for training and also decreases its diversity as well as may suffer from conditions like overfitting training. Hence, an optimum number of training cases are to be taken. Also, from Fig. 6.4 we can see that, after 63.73% training data, the increase in the corresponding accuracy is almost negligible, which can be considered as a saturation point in the training process.

Table 6.5 Classification accuracy of the proposed scheme with varying training and testing data size

Case No.	Number of training cases and its % w.r.t. total number of cases	Number of test cases	Accuracy (%)
1	28,398 (35.53%)	51,522 (64.47%)	98.89
2	36,630 (45.83%)	43,290 (54.17%)	99.22
3	42,024 (52.58%)	37,896 (47.42%)	99.51
4	50,934 (63.73%)	28,986 (36.27%)	99.91
5	55,017 (68.84%)	24,903 (31.16%)	99.91
6	61,425 (76.86%)	18,495 (23.14%)	99.92
7	65,040 (81.38%)	14,880 (18.62%)	99.92
8	69,840 (87.39%)	10,080 (12.61%)	99.93
9	75,720 (94.74%)	4200 (5.26%)	99.93

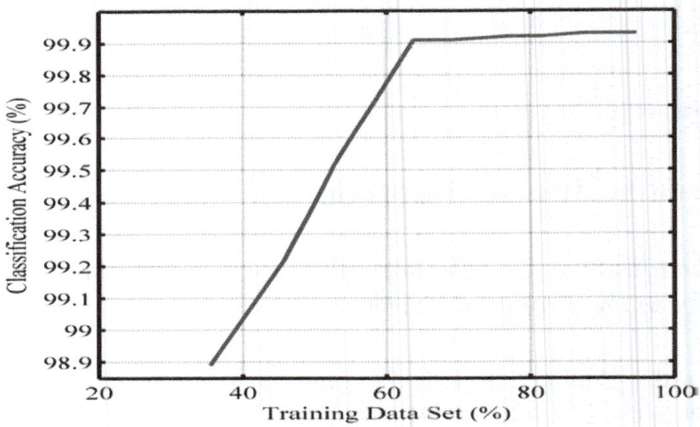

Fig. 6.4 Graph of training data versus percentage accuracy

6.6.2 Classification Accuracy for Various Test Cases

As per the power system fault statistics, approximately 12% of faults take place inside the transformer [34]. Usually, for inter-phase fault, fault resistance is very small and in general, it does not exceed 0.5 Ω. However, fault resistance may be higher during earth fault because of oil and insulation resistance. In the proposed work fault resistance varied up to 15 Ω for internal fault and up to 20 Ω during an external fault in steps of 5 Ω as described in Tables 6.1 and 6.2.

Moreover, as the selection of the size of training data is cleared from Sect. 6.2, the next question may arise, which training data is to be chosen? To verify this, K-fold cross-validation is applied to different training and unseen testing data (Table 6.8). The values of training parameters are chosen randomly along with strictly adhering to the size of training data (63.73% in our case). Table 6.8 shows three cases formed by manually separating the training data from the total available data. From Table 6.8, it is visible that the variation in training data does not affect significantly the accuracy of the algorithm i.e. accuracy remains almost the same in all the 3 test cases. So, this study supports that the accuracy of HE-ELM is independent of the selection of training data. Hence, it can be inferred that even if the unseen data set is applied for testing/validation, the proposed technique gives satisfactory output in terms of higher classification accuracy (average 99.906%).

6.7 Comparison of Proposed Techniques with Existing ELM, SVM and PNN Based Scheme

Recently, SVM and PNN classifier techniques are mainly proposed by various researchers to discriminate fault and abnormal conditions in power systems. The proposed HE-ELM scheme with various parameter variations is compared to shows its effectiveness concerning existing SVM, standard ELM and PNN schemes. Due to the huge numbers of support vectors, SVM becomes a more complicated and time-consuming scheme. The same situation is observed in the PNN technique in terms of testing time.

Table 6.9 shows a comparison of HE-ELM based on accuracy with SVM, PNN and ELM. It is perceived that the offline learning time of HE-ELM is higher than SVM, PNN and ELM scheme while it is to be noted that SVM takes 33 ms as a testing time and having 99.77% accuracy with optimized parameters and 205 support vectors [8]. Moreover, PNN takes 39 ms to classify the same test data with 99.53% accuracy with the best smoothing factor. Similarly, the standard ELM provides 99.83% accuracy for fault classification. Finally, it is to be judged that the proposed HE-ELM technique gives higher classification accuracy (99.91%).

Table 6.6 show classification accuracy among all 28,986 numbers of fault cases. The test cases which correctly classified are denoted as True Positive (TP) and correctly not classified are designated as True Negative (TN). For validation of the proposed scheme total of 28,986 test cases are considered among them 28,959 are accurately classified (TP) and 27 test cases are false classified (TN). As shown in Table 6.6, the proposed algorithm gives an overall 99.91% accuracy which is competent to all existing classifier techniques. It is to be cleared from Table 6.6 that during inrush and external fault conditions proposed scheme discriminate perfectly with 100% accuracy. Internal fault and the external fault with CT saturation are two major issues for consideration unit type protection of transformer. During external fault along with CT saturation conditions, accuracy is also compatible and more than 99.80 percentage in every case. It is inevitably noticed that the proposed HE-ELM algorithm provides better sensitivity for all internal fault and remain stable for all external abnormalities.

Table 6.6 Classification accuracy of the proposed HE-ELM scheme for different fault cases

Sr. No.	Faults cases	Faults/abnormalities	Number of test cases	TP	TN	Accuracy (%)
1	**All types of internal faults**	Turn to turn	2016	1999	17	99.16
		Prim. to sec. winding	4032	4027	5	99.88
		Internal winding	13,440	13,438	2	99.99
2	**External faults**	220 kV bus	1800	1800	0	100.00
		220 kV line	5400	5400	0	100.00
3	**External fault with CT saturation**	220 kV bus	1800	1797	3	99.83
4	**Inrush conditions**	Initial inrush	360	360	0	100.00
		Sympathetic inrush	69	69	0	100.00
		Recovery inrush	69	69	0	100.00
	Total data	**79,920**	**28,986**	**28,959**	**27**	**99.91**

Bold represents a number of data/parameter used for the validation of the proposed scheme

Table 6.7 shows fault type-wise classification accuracy of the HE-ELM algorithm. It clearly shows that with 13,438 TP and 02 TN cases, proposed HE-ELM gives 99.99% classification accuracy under internal winding fault among 13,440 total test data. It also gives 99.97% classification accuracy during external fault with and without CT saturation conditions with 8997 TP and 03 TN among 9000 total test data. The algorithm provides 99.95% accuracy during L-G internal fault which is likely to occur in the transformer. This shows the effectiveness of the Proposed HE-ELM algorithm. Moreover, during L-G external fault, it provides 100% accuracy. This points out that the proposed scheme is more vigorous and remains inoperative for all major external disturbances (Tables 6.8 and 6.9).

Table 6.7 Fault category wise classification accuracy using HE-ELM

Sr. No.	Fault type	Internal winding fault				External fault with and without CT saturation			
		Total	TP	TN	Accuracy (%)	Total	TP	TN	Accuracy (%)
1	**L-g**	4032	4030	2	99.95	2700	2700	0	100
2	**L-L**	4032	4032	0	100	2700	2700	0	100
3	**L-L-g**	4032	4032	0	100	2700	2700	0	100
4	**L-L-L**	1344	1344	0	100	900	897	3	99.67
	Total	**13,440**	**13,438**	**2**	**99.99**	**9000**	**8997**	**3**	**99.97**

Bold represents a number of data/parameter used for the validation of the proposed scheme

Table 6.8 Cross-validation of the proposed scheme for different training and testing data

Total data (79,920) as shown in Table 6.1			Fault type	Source impedance %	Fault location in %	FIA/switching instant	Fault resistance (£1)	Load angle (S)	Residualflux (%)	Trainin g data (50,934)	Testing data (28,986)	Accuracy (%)
Case-1	Internal (48,720)	Turn to turn fault	6 (3 on primary and 3 on secondary)	3 (80, 100, 120)	6 (0.2, 1, 2, 3, 4, 7)	7 (0°, 15°, 30°, 60°, 120°, 135°, 150°)	–	4 (0°, 5°, 15 20°)	–			
		Inter winding	3 (all in 3 winding)	3 (80, 100, 120)	6 (0, 15, 30, 60, 75, 90)	7 (0°, 15°, 30°, 60°, 120°, 135°, 150°)	4 (0, 5, 10, 15)	4 (0°, 5°, 15°, 20°)	–	29,232	19,488	
		Internal winding	10 (L-G, LL, LL-G, LLL)	3 (80, 100, 120)	6 (0, 30, 45, 60, 75, 90)	7 (0°, 15°, 60°, 90°, 120°, 135°, 150°)	4 (0, 5, 10, 15)	4 (0°, 5°, 15°, 20°)	–			
	External (30,000)	Ext. with and without CT saturation	10 (L-G, LL, LL-G, LLL)	3 (80, 100, 120)	–	7 (0°, 15°, 60°, 90°, 120°, 135°, 150°)	5 (0, 5, 10, 15, 20)	4 (0°, 5°, 15°, 20°)	–	21,000	9000	
		Fault on 220 kV line	10 (L-G, LL, LL-G, LLL)	3 (80, 100, 120)	3 (1 km, 20 km, 15 km)	7 (0°, 15°, 60°, 90°, 120°, 135°, 150")	5 (0, 5, 10, 15, 20)	4 (0°, 5°, 15°, 20°)	–			99.91
	Inrush (1200)	Initial inrush	–	3 (80, 100, 120)	–	9 (0°, 15°, 30°, 45°, 60°, 90°, 120°, 135°, 150°)	–	4 (0°, 5°, 15°, 20°)	5 (0, 10, 45, 60, 80)			

(continued)

Table 6.8 (continued)

Total data (79,920) as shown in Table 6.1			Fault type	Source impedance %	Fault location in %	FIA/switching instant	Fault resistance (£1)	Load angle (S)	Residualflux (%)	Trainin g data (50,934)	Testing data (28,986)	Accuracy (%)
		Sympathetic inrush	–	3 (80, 100, 120)	–	9 (0°, 15°, 30°, 45°, 60°, 90°, 120°, 135°, 150°)	–	3 (0°, 15°, 20°)	–	702	498	
		Recovery inrush	–	3 (80, 100, 120)	–	9 (0°, 15°, 30°, 45°, 60°, 90°, 120°, 135°, 150°)	–	3 (0°, 15°, 20°)	–			
Case-2	Internal (48,720)	Turn to turn fault	6 (3 on primary and 3 on secondary)	3 (80,100, 120)	6 (0.2, 1, 2, 3, 5, 7)	7 (0°, 15°, 30°, 90°, 120°, 135°, 150°)	–	4 (0°, 10°, 15°, 20°)	–			
		Inter winding	3 (all in 3 winding)	3 (80, 100, 120)	6 (0, 15, 45, 60, 75, 90)	7 (0°, 15°, 30°, 90°, 120°, 135°, 150°)	4 (0, 5, 10, 15)	4 (0°, 10°, 15°, 20°)	–	29,232	19,488	
		Internal winding	10 (L-G, LL, LL-G, LLL)	3 (80,100, 120)	6 (0, 15, 45, 60, 75, 90)	7 (0°, 15°, 30°, 90°, 120°, 135°, 150°)	4 (0, 5, 10, 15)	4 (0°, 10°, 15°, 20°)	–			
	External (30,000)	External with and without CT saturation	10 (L-G, LL, LL-G, LLL)	3 (80, 100, 120)	–	7 (0°, 15°, 30°, 90°, 120°, 135°, 150°)	5 (0, 5, 10, 15, 20)	4 (0°, 5°, 10°, 20°)	–	21,000	9000	

(continued)

Table 6.8 (continued)

Total data (79,920) as shown in Table 6.1			Fault type	Source impedance %	Fault location in %	FIA/switching instant	Fault resistance (£1)	Load angle (S)	Residualflux (%)	Trainin g data (50,934)	Testing data (28,986)	Accuracy (%)
		Fault on 220 kV line	10 (L-G, LL, LL G, LLL)	3 (80, 100, 120)	3 (1 km, 20 km, 15 km)	7 (0°, 15°, 30°, 90°, 120°, 135”, 150°)	5 (0, 5, 10, 15, 20)	4 (0°, 5°, 10°, 20°)	–			99.89
	Inrush (1200)	Initial inrush	–	3 (80, 100, 120)	–	9 (0°, 15°, 30°, 45°, 60°, 75°, 120°, 135°, 150°)	–	4 (0°, 5°, 10°, 20°)	5 (0, 10, 25, 60, 80)			
		Sympathetic inrush	–	3 (80, 100, 120)	–	9 (0°, 15°, 30°, 45°, 60°, 75°, 120°, 135°, 150°)	–	3 (0°, 5°, 20°)	–	702	498	
		Recovery inrush	–	3 (80, 100, 120)	–	9 (0°, 15°, 30°, 45°, 60°, 75°, 120°, 135°, 150°)	–	3 (0°, 5°, 20°)	–			
Case-3	Internal (48,720)	Turn to turn fault	6 (3 on primary and 3 on secondary)	3 (80, 100, 120)	6 (0.2, 1, 2, 3, 4, 5)	7 (0°, 15°, 30°, 60°, 90°, 120°, 135°)	–	4 (0°, 5°, 10°, 15°)	–			
		Inter winding	3 (all in 3 winding)	3 (80, 100, 120)	6 (0, 15, 30, 45, 60, 75)	7 (0°, 15°, 30°, 60°, 90°, 120°, 135°)	4 (0, 5, 10, 15)	4 (0°, 5°, 10°, 15°)	–	29,232	19,488	

(continued)

Table 6.8 (continued)

Total data (79,920) as shown in Table 6.1		Fault type	Source impedance %	Fault location in %	FIA/switching instant	Fault resistance (£1)	Load angle (S)	Residualflux (%)	Trainin g data (50,934)	Testing data (28,986)	Accuracy (%)
	Internal winding	10 (L-G, LL, LL-G, LLL)	3 (80, 100, 120)	6 (0, 15, 30, 45, 60, 75)	7 (0°, 15°, 30°, 60°, 90°, 120°, 135°)	4 (0, 5, 10, 15)	4 (0°, 5°, 10 15°)	–			
External (30,000)	External with and without CT saturation	10 (L-G, LL, LL-G, LLL)	3 (80, 100, 120)	–	7 (0°, 15°, 30°, 60°, 90°, 120°, 135°)	5 (0, 5, 10, 15, 20)	4 (0°, 5°, 10 15°)	–	21,000	9000	99.91
	Fault on 220 kV line	10 (L-G, LL, LL-G, LLL)	3 (80, 100, 120)	3 (1 km, 20 km, 15 km)	7 (0°, 15°, 30°, 60°, 90°, 120°, 135°)	5 (0, 5, 10, 15, 20)	4 (0°, 5°, 10 15°)	–			
Inrush (1200)	Initial inrush	–	3 (80, 100, 120)	–	9 (0°, 15°, 30°, 45°, 60°, 75°, 90°, 120°, 135°)	–	4 (0°, 5°, 10°, 15°)	5 (0, 10, 25, 45, 60)			
	Sympathetic inrush	–	3 (80, 100, 120)	–	9 (0°, 15°, 30°, 45°, 60°, 75°, 90°, 120°, 135°)	–	3 (0°, 5°, 15°)		702	498	
	Recovery inrush	–	3 (80, 100, 120)	–	9 (0°, 15°, 30°, 45°, 60°, 75°, 90°, 120°, 135°)	–	3 (0°, 5°, 1 5°)	–			

Table 6.9 Comparisons of the proposed HE-ELM scheme with SVM and PNN scheme

Sr. No.	Faults cases/abnormalities	Total test cases	Proposed scheme (HE-ELM)			SVM scheme			PNN scheme			ELM scheme		
			TP	TN	%η	TP	TN	%η	TP	TN	%η	TP	TN	%η
1	Internal faults	19,488	19,464	24	99.88	19,447	41	99.79	19,409	79	99.59	19,453	35	99.82
2	External faults	7200	7200	00	100	7186	14	99.80	7175	25	99.65	7192	08	99.89
3	The external fault with CT saturation	1800	1797	03	99.83	1792	08	99.55	1779	21	98.83	1795	05	99.72
4	Inrush conditions	498	498	00	100	494	04	99.19	487	11	97.79	497	01	99.80
	Total	28,986	28,959	27	**99.91**	28,919	67	**99.77**	28,850	136	**99.53**	28,937	49	**99.83**

TP—True positive, TN—True negative, %η—Percentage accuracy

6.8 Hardware Setup and Test Results

Figure 6.5 shows hardware prototype set up in the laboratory environment for 50 kVA, 440/220 V transformer protection to validate the proposed algorithm. Also, rheostats and inductors are placed before and after the transformer to replicate the effect of the transmission line which is present in the real-time condition. The transformer's primary side is connected with 3-phase, 440-V separate generator, and the secondary side is connected to 3-phase 220 V electricity board variable supply through simulated transmission lines. Two 6-pole contactors (circuit breakers) are used to connect the transformer with generator and variable utility supply. A set of protective CTs is connected on the primary and secondary side with 25/1 and 50/1 A ratings respectively. Primary side and secondary side internal faults, as well as external faults, are generated by connecting 12 A, 18 Ω variable resistors (rheostats) in fault path. Fault resistors are variable, so adjustable fault current can flow as per the requirement.

The transformer is specially designed in such a way that internal and inter-turn fault can be created through tapping at a different percentage of winding as shown in the below images. The primary side of the transformer is tapped as 254–228–204–180–0 Volts/phase and the secondary side is segmented as 127–114–102–90–0 Volts/phase. Through these tapping internal faults as well as inter-turn faults are possible. During the generation of various internal faults, transformer turns are manually changed (fault location) for all 4 types of faults (1-L-g, 1-LL, 1-LL-g, and 1-LLL). During the said internal fault formation the inserted rheostat in the fault path will be varied to make an effect of low resistance to high resistance (R_f) internal fault in the transformer.

Further extended view of hardware set up is as shown in Figs. 6.6 and 6.7. Figure 6.6 gives the exact detail of the panel with its control diagram of primary as well as the secondary side of the transformer protection panel. Figure 6.7a–c

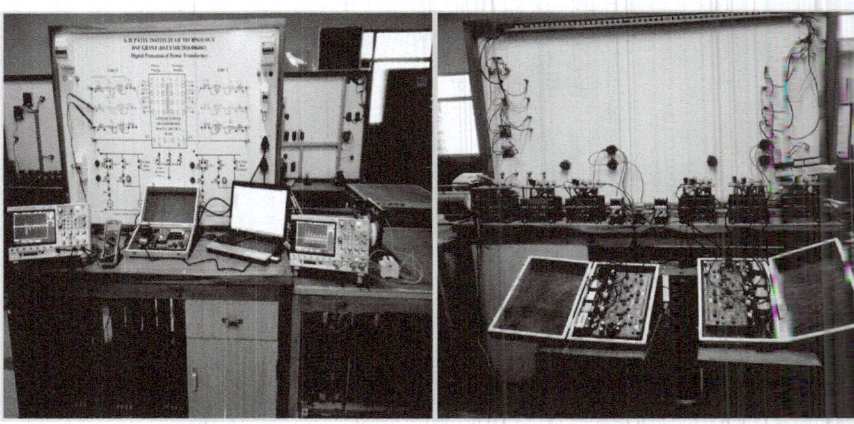

(a) front view (b) rear view

Fig. 6.5 Hardware prototype in laboratory **a** front view, **b** rear view of the panel

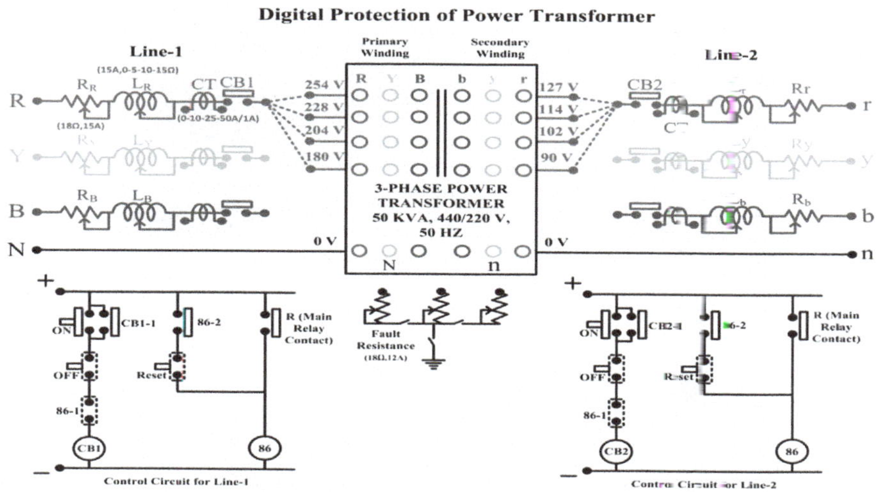

Fig. 6.6 Three phase diagram (with control diagram) for hardware set up to create fault and abnormalities on considered power transformer

gives transformer tapping, placing of variable rheostats and variable inductors to create an equivalent line of the section of the power system. We incorporated the quantitative details of fault created in hardware setup and also included three-phase current waveforms in Figs. 6.9 and 6.18.

External faults are simulated on a series combination of variable rheostats and variable inductors placed on the primary side and secondary side of a transformer outside the CT locations (below images). Like internal faults, during the formation of external fault, fault types, fault location on simulated lines and fault resistances are varied. Moreover, during a certain external fault, 250 Ω rheostats inserted in the secondary of CTs are varied to make the effect of CT saturation. Three separate switches are used to simulate internal faults and another three separate switches are used to create external ground fault conditions respectively.

Similarly, various other switches are incorporated in hardware to create line-line, line-line to the ground, and triple line faults. Ratings and specifications of all these components including fault switches are selected as per the power transfer capacity and fault sustain capacity of that components. The step by step procedure of implementing the proposed algorithm in the CORTEX M4 microcontroller is narrated as below.

1. The μ Vision Integrated Development Environment (IDE) is a powerful platform used to build, edit, and debug the program. With this IDE platform, equations of DWT and proposed fault classifiers are coded in C++ language step by step as mentioned in Sect. 5.3.
2. The program is then executed and compiled in Keil Version 5 software, which will convert the program level language (C++) into machine level language (.HEX).

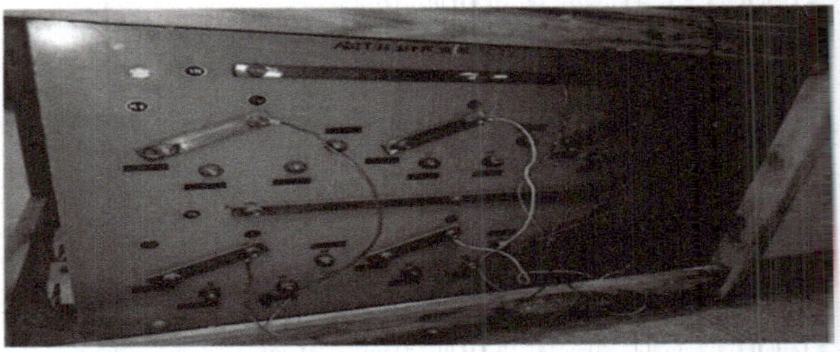

(a) Tapping at Different Percentage of Winding on Primary and Secondary

(b) Six Variable Rheostats to Create Line

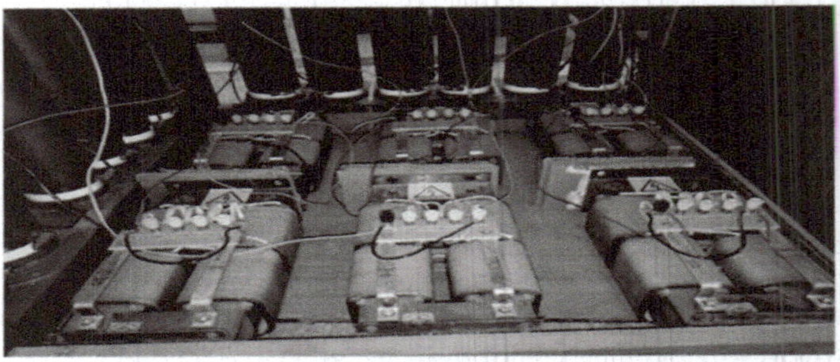

(c) Six Variable Inductors to Create Line

Fig. 6.7 Detailed view of laboratory setup

3. The program is then exported to the CORTEX-M4 microcontroller (STM32F407).
4. After successful uploading of the program in the CORTEX-M4 microcontroller, the post-disturbance current signals from current sensors (scaled-down value after I-to-V conversion) are fed as input to on-chip Analog to Digital Converter (ADC) of the controller.

5. The digital data is then assigned to the variables inside the program which is further applied to execute the main program. The microcontroller will extract the necessary features from the given electrical signal for one cycle post-disturbance data.
6. The next step is to train the HE-ELM model using DWT extracted features of fault/load/inrush data. The training of the HE-ELM is done offline using extracted post fault as well as normal data set (buffered in controller memory).
7. The trained HE-ELM model is now ready for testing of next real-time disturbance/abnormal condition and capable to judge whether it is an internal fault or not?
8. If the Learning Machine detects it as a transformer internal fault then the microcontroller will send a trip signal to its output port which is further applied to an external solid-state actuator circuit (electronic relay).

Various fault cases are tested on the trained algorithm to authenticate the performance of the proposed scheme on a real-time basis. The high pulse (+5 V) output of HE-ELM on the microcontroller board denotes internal fault and low pulse (0 V) denotes external fault/inrush conditions.

Table 6.10 gives information about various test conditions performed on hardware setup in the laboratory. Total 91 fault cases of real-time data (17 inrushes + 37 internal fault + 37 external faults including CT saturation effect) are performed on the hardware setup. From which, 55 cases as training data and the remaining 36 cases as testing data are utilized for validation of the proposed algorithm in a laboratory environment.

In case of any fault occurs inside the transformer, the proposed HE-ELM detects the fault condition and can send trip signal. Contrary, if the case is of inrush condition or external abnormalities, the algorithm can sense the fault condition and will remain stable. Internal faults [{4 (Fault Location) * 4 (F_{Type}) * 2 (R_f)} + 5 (TTF)] and external faults [{4 (Fault Location) * 4 (F_{Type}) * 2 (R_f)} + 5 (CT saturation)] are simulated on the developed hardware.

Figure 6.8a–e show the hardware waveform captured by digital storage oscilloscope (DSO) during inrush, internal fault condition (L-G), internal fault condition (LL-G), external fault condition and CT saturation condition during external fault respectively.

It is to be noted from Table 6.10 that classification accuracy obtained by the HE-ELM algorithm is 100% during validation. Whereas the fault classification accuracy gained by SVM is 97.22% and on the other hand the PNN scheme gives 94.44% accuracy in the hardware validation. While standard ELM gives the classification of 100%. SVM and PNN false classify one case and two cases respectively out of a total of 36 cases.

Table 6.10 Fault data generation through hardware setup

Fault cases	Inrush data	Training data (inrush)	Internal fault data	Training data (internal)	External fault data	Training data (external)
Inrush at different inception angle	17	11	–	–	–	–
Turn to turn	–	–	+05	+04	–	–
Fault location	–	–	04	03	04	03
Fault type	–	–	04	03	04	03
Fault resistance (*Rf*)	–	–	02	02	02	02
CT saturation during fault	–	–			+05	+04
Total and training data	17	11	37	22	37	22
Testing data	06		15		15	
HE-ELM	100% (TT = 27 ms) (36 TP/36 total)					
SVM	97.22% (TT = 33 ms) (35 TP/36 total)					
PNN	94.44% (TT = 39 ms) (34 TP/36 total)					
ELM	100% (TT = 34 ms) (36 TP/36 total)					

TT—Testing Time, ms—millisecond, TP—True Positive

6.9 Additional Tested DSO Results

Moreover, few more hardware validation figures are provided below for reference. These waveforms are directly captured from DSO.

The waveform shown in Fig. 6.9 is for inrush condition of the transformer. During this condition secondary side is kept open and then the inrush waveform is fetched from the primary side of the transformer which is the actual scenario of the real field condition.

The Fig. 6.10 below is captured by DSO when a single line to ground fault is created on the tapping of the transformer to ground through a high resistant of 18 Ω.

Figure 6.11 shown above is for the waveform for line to ground (L-G) fault created in transformer internal winding with low fault resistance in fault path. Hence the magnitude of fault current is higher than that shown in previous Fig. 6.10.

Figure 6.12 shows the current waveform during internal fault (L-G with slight decaying DC component) condition. This type of condition is taking place in the

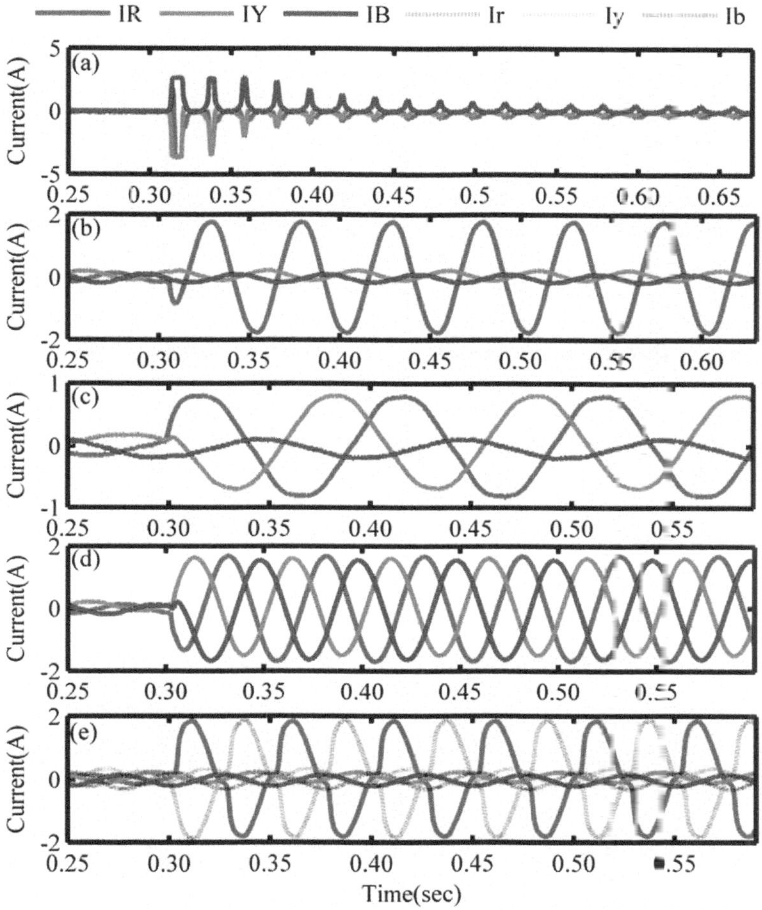

Fig. 6.8 Transformer primary and secondary side current waveform for case **a** Inrush **b** internal fault (L-G) **c** internal fault (LLg) **d** external fault (LLL) **e** external fault (L-G) with CT saturation

transformer while the fault is of inductive in nature. The below waveform shows replica of the fault condition which is taking place in the real field during transformer operation.

Also, Fig. 6.13 shown here displays the three phase waveform captured during double line to ground (LL-G) fault case. It is clearly seen from the waveform that two faulted phase current magnitude is increased after inception of internal fault. Before the fault condition taking place the waveform are symmetrical to each other and having equal current magnitude (load).

Figure 6.14 shows the waveform of double line (LL) fault created inside the transformer protection zone.

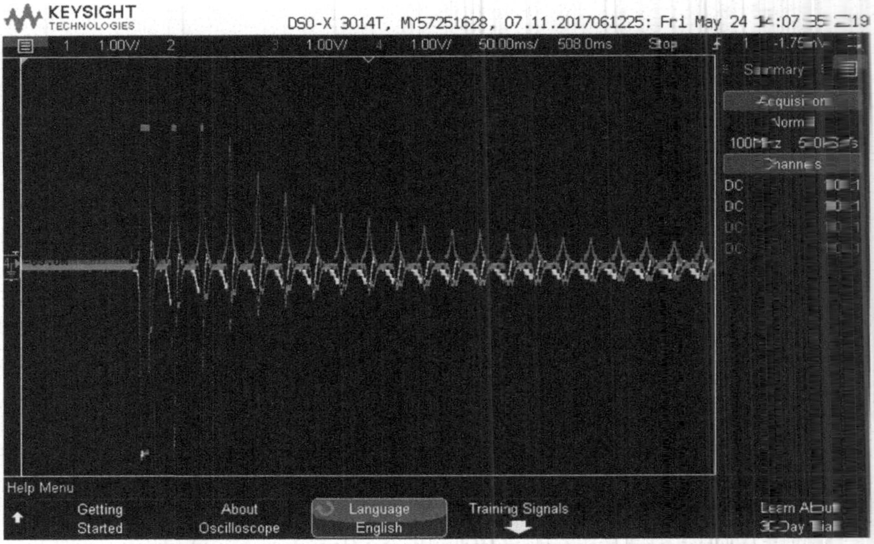

Fig. 6.9 Transformer primary side current waveforms for inrush condition

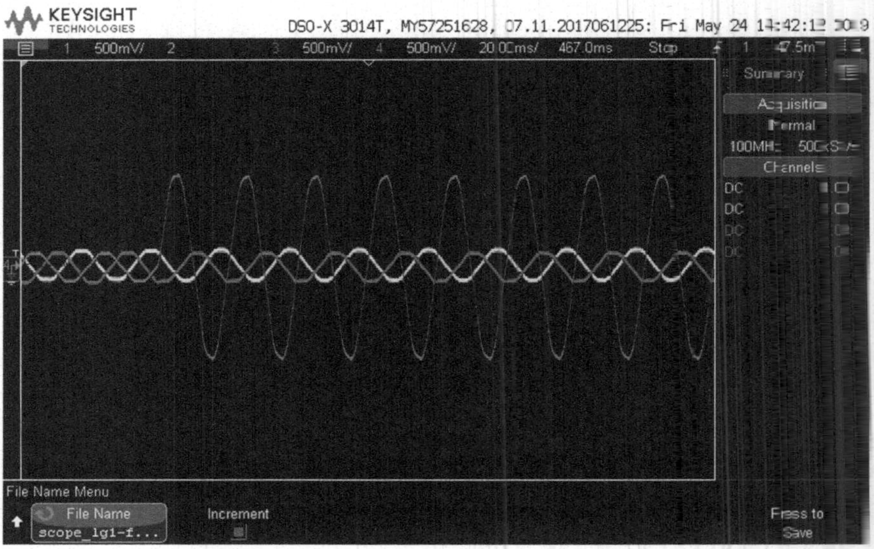

Fig. 6.10 Transformer primary side current waveforms for internal (L-G) fault condition

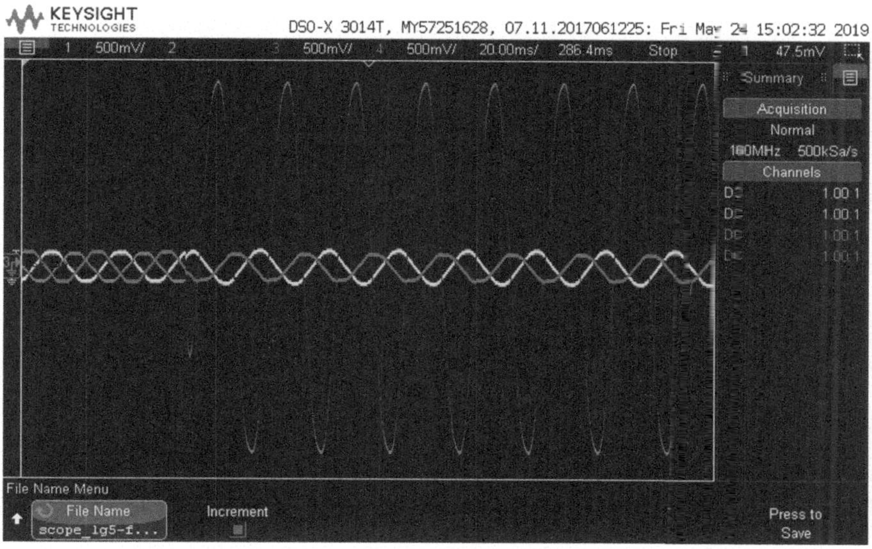

Fig. 6.11 Transformer primary side current waveforms for internal (L-G) fault condition with low fault resistance

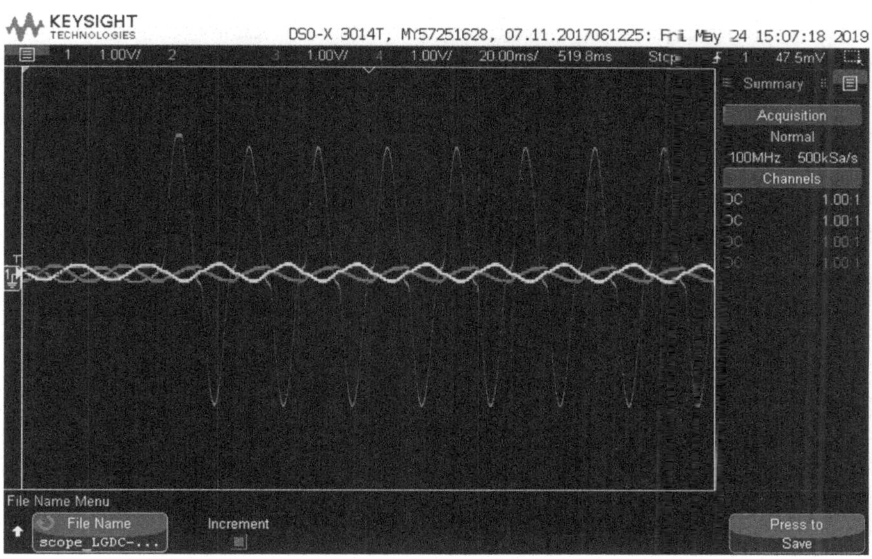

Fig. 6.12 Transformer primary side current waveforms for internal fault condition (L-G fault with slight decaying DC component)

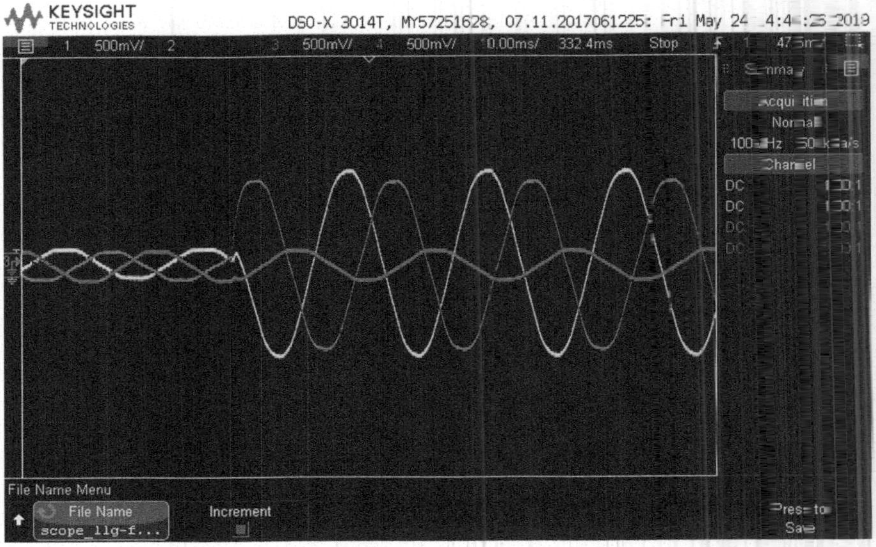

Fig. 6.13 Transformer primary side current waveforms for internal (LL-G) fault condition

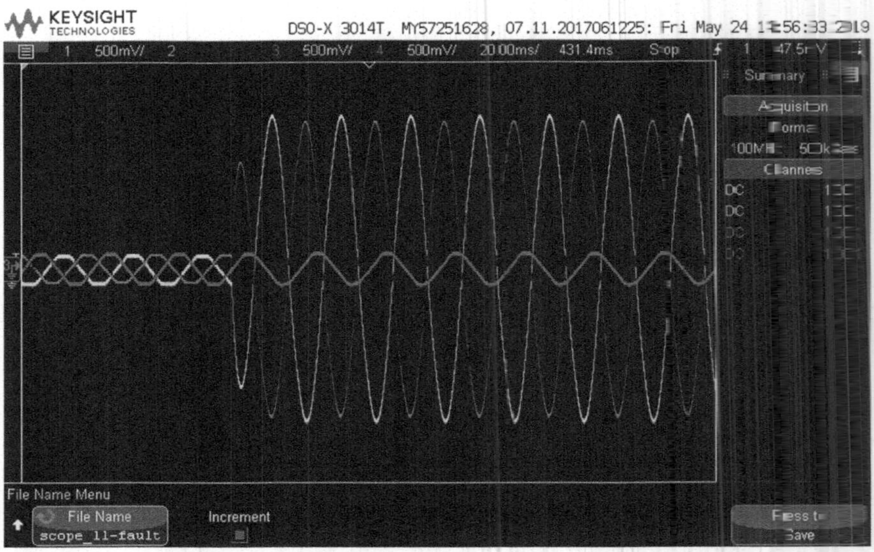

Fig. 6.14 Transformer primary side current waveforms for internal (LL) fault condition

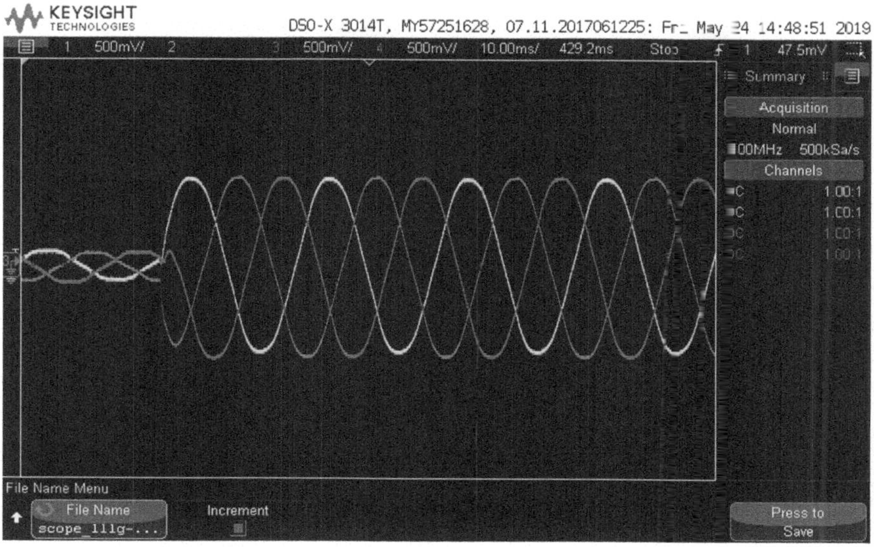

Fig. 6.15 Transformer primary side current waveforms for internal (LLL) fault condition on lower tapping

Figure 6.15 illustrated below replicates the waveform condition during triple line (LLL) fault generated in the transformer at lower tapping towards neutral. From the Fig. 6.13, we can see that the waveform is initially symmetrical to each other. However, after fault inception the magnitude of fault current only increases but symmetricity will remain same.

The same condition can be shown in Fig. 6.16 for internal LLL fault case simulated on higher tapping towards terminal of the transformer.

Figure 6.17 shows the current waveform for external fault condition for single line to ground (L-G) fault. Here the current is measured from primary side of the transformer. While Fig. 6.18 depicts current waveform measured from secondary side of the transformer for the same faulty condition (i.e. L-G).

6.10 Benefits of the Proposed Scheme

The analysis presented in the previous section and results of the proposed HE-ELM scheme emphasizes various benefits compared to existing schemes, which are narrated as below.

- HE-ELM is not more sensitive for hidden nodes unlike other gradient-based learner algorithms (PNN), whereas SVM requires the setting of margin trade-off and regularization parameters.

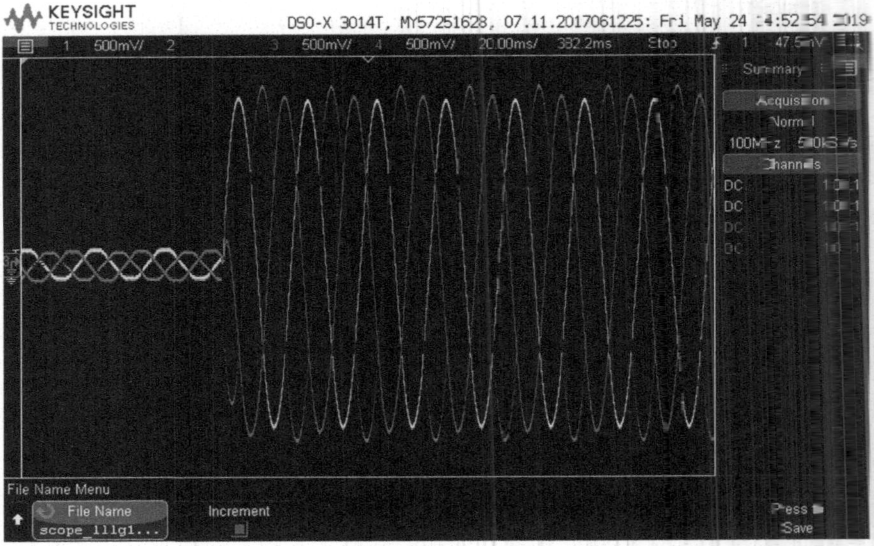

Fig. 6.16 Transformer primary side current waveforms for internal (LLL) fault condition on higher tapping

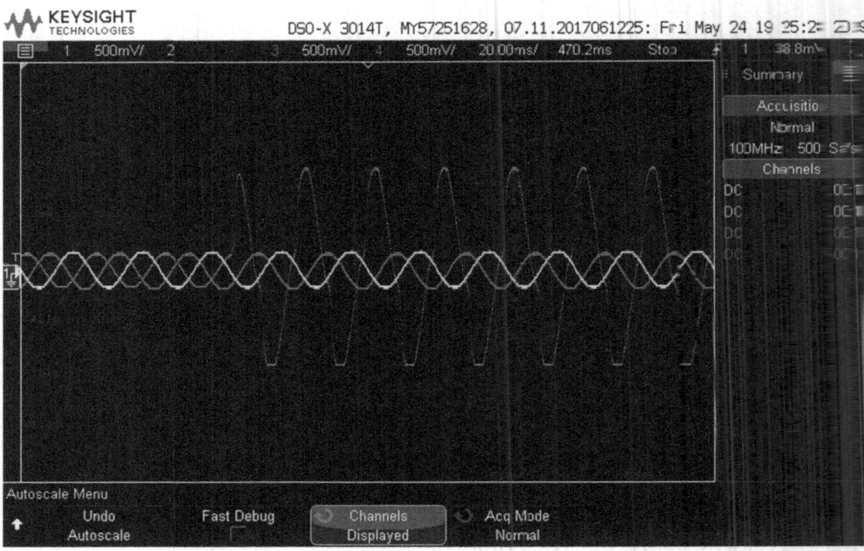

Fig. 6.17 Transformer primary side current waveforms for external (L-G) fault condition

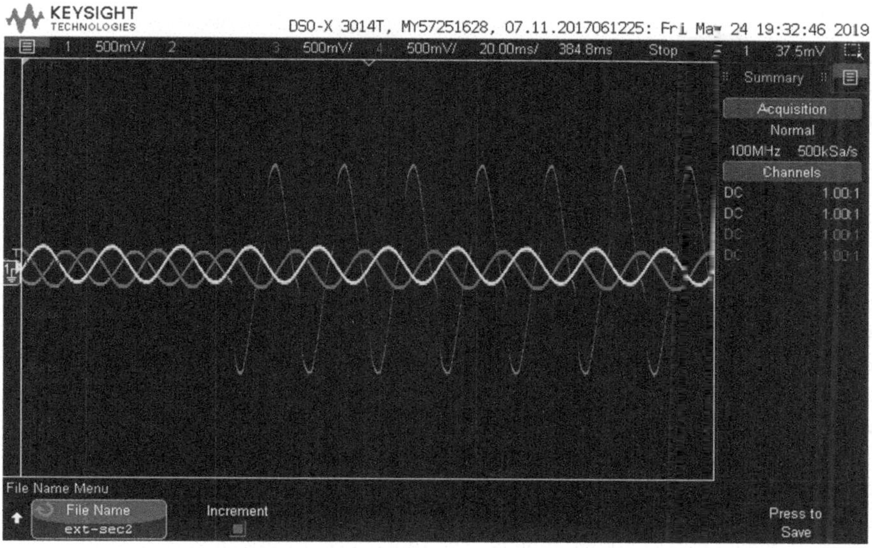

Fig. 6.18 Transformer secondary side current waveforms for external (L-G) fault condition

- HE-ELM gives very stable operation and efficient performance under, use of random nodes with hidden layers.
- Under the noisy environment also, HE-ELM performs better than conventional classifier techniques. Hence, it may not require pre-processing of data (current signals) every time.
- Batch learning kernel solution of HE-ELM is much simpler than other kernel learning algorithms such as LS-SVM.
- The performance of SVM, PNN and ELM for unknown feature vector is slower than HE-ELM.
- Table 6.10 reveals that the proposed HE-ELM algorithm outperforms compare to SVM, PNN and ELM in terms of classification accuracy and speed of fault discrimination.
- **The novelty of this work with respect to previous RVM techniques (Refer to Ch-5)**.

We have utilized the technique of HE-ELM proposed by Cai et al. [20] with slight modification, as per our requirement in this research work. We are not claiming that it is our invention, we have only utilized the concept of HE-ELM [20] with modification in steps of an algorithm developed for the proposed technique. From the refereed article, the authors found that the HE-ELM learning machine may help positively to provide the discrimination between in zone, out of zone fault, and inrush condition in the transformer. Also, it may capable enough to tackle huge data sets, as the higher size of training data sets will help the protective algorithm to perform fine and hassle-free discrimination between internal fault conditions and various other abnormalities. The higher the input size chosen optimally, the higher will be the accuracy and reliability of the protective scheme. Hence, we have decided to utilize the learning ability of the HE-ELM and applied it to develop a sound protective scheme for power transformer. Consequently, the simulation and hardware results obtained prove the efficacy of the HE-ELM in terms of operating time and accuracy.

As stated, we have utilized the competency of the HE-ELM in our research work to successfully discriminate internal fault and other abnormalities. The novelty of the paper we can give in terms of application in power transformer unit protection by generating various faults and abnormalities in and outside of the considered transformer. Further, various data generated will provide a high level of training data sets, which will be utilized to design a highly promising result-oriented protection scheme for power transformer. By doing a thorough review we have found that many researchers had already successfully worked in this field even though there remains scope of improvement as stated in the introduction. Further, to prevent undesired tripping of power transformer from outside abnormal conditions, classifier, and waveform pattern recognition based techniques are trained using vast training data sets. The huge training data set is generated here because in the power system there is vast variation occurs in system parameters. The root causes of this failure in developing a sound protection scheme can be given as bulky training as well as testing data, which ultimately causes slow response during a fault condition, insufficient training data which may ultimately result in no operation of the relay, bulky hidden layer which requires more training time. To tackle these situations, HE-ELM along with feature bagging and Bernoulli's probability has been incorporated, which will optimally choose hidden connections between and also reduce the over-fitting of the learning machine. Hence, we have found that the HE-ELM works batter for bulky training data and also provides desired classification accuracy with optimized learning parameters which will help, to develop a sound transformer protection scheme. Moreover, hardware setup which is a replica of the real field has been built which will help to validate the algorithm correctly as it provides a reflection of transmission lines as well as transformer inherent characteristics. A high-speed CORTEX M4 ARM processor has been programmed and incorporated in the hardware testing which has completely authenticate the protective scheme against various real-time faults and abnormalities. The capability of the scheme can be seen from the derived algorithm as well as hardware prototype results. The article includes accuracy and operational time comparison of the proposed protection scheme with various modern protective schemes which gives ready reference to the stack holders about the effectiveness of the proposed HE-ELM algorithm over other methods. Moreover, a comparison of variation in size of training and testing data sets has been incorporated with variation in the selection of training data while keeping training data set of constant size. The obtained learning curve is also helpful in finding the variation in accuracy concerning the size of the data set.

Nowadays, transformer protection demands fast fault clearing (minimum time) and hence, avoids problems of transient stability. At the same time, unnecessary operation of the transformer protection scheme results in outage as well as raises stability problems. Hence, the power system engineer must achieve discrimination between the desired tripping of the transformer protection scheme during in-zone fault and inhibition of the same scheme during out of zone fault. Moreover, there is always a compromise between maximum protection functions available/incorporated in the scheme and minimum cost. Hence, a particular scheme available in the field is unable to protect the power transformer during verities of fault conditions. As the external parameters in power system changes, the real-time field data are required to retrain/reuse the classifier algorithm. HE-ELM is not required to be trained online in real-time. It is to be noted that the percentage fault discrimination accuracy provided by the proposed HE-ELM based algorithm is within the acceptable limit in case of the unseen data set (test data). Therefore, there is no need to retrain the HE-ELM during little change in the power system network. However, it is necessary to re-train the HE-ELM during a major change in the power system network. In case, requirement arises for retraining with newly added training cases, the training (previous training) can be performed off-line with an updated training dataset and the HE-ELM algorithm can be updated online in hardware module without taking the relay out of service.

Further, concerning the article "Design and development of fault classification algorithm based on relevance vector machine for power transformer" [8], we agree that the test cases we have considered in both the paper look similar. For faults and abnormalities generation, factors that mostly affects the power transformer protection scheme are described for the validation. Variation in system parameters during different situations in power systems like the type of fault, source impedance, fault location, fault inception angle, the magnitude of fault resistance, load angle, circuit breaker closing time, residual flux, etc. are taken into consideration. The similarity is seemed in these proposed and above-mentioned paper is only because we want to take as many parameters and variation that affects power transformer protection and want to provide complete protection to the power transformer from all the angle and perspective. We have varied all these above-mentioned parameters simultaneously in a batch by utilizing the multi-run feature of PSCAD^{TM} software. We have considered more parameter variation than the previous one. Although we can consider more parameters other than these which adds on merit in the research but which will make the article more lengthy and complicated. Hence, all possible parameter variation and test cases that affect most to power transformer protection are considered for both the papers. Moreover, the validation and classification method of test cases in both the papers are different in terms of variation in training and testing data set as well as for unseen data. The above exercise is only carried out in HE-ELM based article and not in the previous article based on the relevance vector machine. In a previously published paper of RVM, only software is used to validate the algorithm through data generated on the prototype. Whereas, in this article, the ARM cortex M4 high-speed processor is physically used to check the real-time application of the developed scheme. Moreover, the proposed HEELM algorithm is tested on a three-phase transformer whereas our previously published RVM based article has been tested on a single-phase transformer only.

6.11 Summary

This research article presents HE-ELM and DWT based new protective scheme for the transformer in the zone and out of zone fault classification. Various fault cases are generated by PSCAD software with a multi-run block. One cycle post fault current signals are acquired from both sides of the transformer considering 4 kHz sampling frequency. The proposed HE-ELM classifier is validated for various test cases like an internal fault, external fault, external fault with CT saturation, and inrush conditions in the transformer. Out of 79,920 cases, 28,986 (36.26%) have been considered for validation of the proposed HE-ELM algorithm. It turns out from result analysis that HE-ELM outperforms than other classifier techniques like SVM, PNN and ELM. It is observed from results that HE-ELM based classifier techniques provide higher classification accuracy more than 99 percent with minimum validation time of 27 ms for transformer protection. Additionally, it provides better security against inrush and external fault condition even in CT saturation conditions. The justification behind the selection of the size of the training data set and the type of training data set is also specified in the validation section. A hardware prototype is also designed by certainly considering all the real-time fault situations that are present in the transformer to authenticate the developed algorithm. The developed algorithm is tested using the CORTEX-M4 microcontroller. A total of 91 fault cases have been generated out of which 36 data are utilized for testing purposes. It is observed from the experimental performance that HE-ELM outperformed for hardware test results by providing 100% classification accuracy. Thus, the proposed HE-ELM based technique is more efficient in the practical field due to faster fault classification with higher performance.

6.12 Published Article Based on This Work

M. B. Raichura, N. G. Chothani, and D. D. Patel, “Identification of internal fault against external abnormalities in power transformer using hierarchical ensemble extreme learning machine technique,” IET Sci. Meas. Technol., vol. 14, no. 1, pp. 111–121, 2020.

References

1. Shah AM, Bhalja BR (2013) Discrimination between internal faults and other disturbance in transformer using the support vector machine-based protection scheme. IEEE Trans Power Deliv 28(3):1508–1515
2. Huang G-B, Zhu Q-Y, Siew C-K (2006) Extreme learning machine: theory and applications. Neurocomputing 70(1):489–501
3. Chorowski J, Wang J, Zurada JM (2014) Review and performance comparison of SVM- and ELM-based classifiers. Neurocomputing 128(Supplement C):507–516

4. Bucurica M, Dogaru M, Dogaru I (2015) A comparison of extreme learning machine and support vector machine classifiers. In: 2015 IEEE international conference on intelligent computer communication and processing (ICCP), pp 471–474
5. Koroglu S, Demircali A (2016) Diagnosis of power transformer faults based on multi-layer support vector machine hybridized with optimization methods. Electr Power Compon Syst 44(19):2172–2184
6. Paydarnia H, Hajiaghasi S, Abbaszadeh K (2014) Improved structure of PNN using PCA in transformer fault diagnostic. Arab J Sci Eng 39(6):4845–4851
7. Balaga H, Gupta N, Vishwakarma DN (2015) GA trained parallel hidden layered ANN based differential protection of three phase power transformer. Int J Electr Power Energy Syst 67:286–297
8. Patel D, Chothani NG, Mistry KD, Raichura M (2018) Design and development of fault classification algorithm based on relevance vector machine for power transformer. IET Electr Power Appl 12(4):557–565
9. Maya P, Roopasree K, Soman KP (2015) Discrimination of internal fault current and inrush current in a power transformer using empirical wavelet transform. Procedia Technol 21:514–519
10. Chen J, Phung T, Blackburn T, Ambikairajah E, Zhang D (2016) Detection of high impedance faults using current transformers for sensing and identification based on features extracted using wavelet transform. IET Gener Transm Distrib 10(12):2990–2998
11. Shah AM, Bhalja BR, Patel RM (2018) New protection scheme for power transformer based on superimposed differential current. IET Gener Transm Distrib 12(14):3587–3595
12. Lin X, Lu J, Zhang R, Tong N, Adio OS, Li Z (2015) Internal fault fast identification criterion based on superimposed component comparison for power transformer. Int J Electr Power Energy Syst 73:491–497
13. Ghanbari T, Samet H, Ghafourifard J (2016) New approach to improve sensitivity of differential and restricted earth fault protections for industrial transformers. IET Gener Transm Distrib 10(6):1486–1494
14. Dashti H, Davarpanah M, Sanaye-Pasand M, Lesani H (2016) Discriminating transformer large inrush currents from fault currents. Int J Electr Power Energy Syst 75:74–82
15. Medeiros RP, Costa FB (2018) A wavelet-based transformer differential protection with differential current transformer saturation and cross-country fault detection. IEEE Trans Power Deliv 33(2):789–799
16. Medeiros RP, Costa FB (2018) A wavelet-based transformer differential protection: internal fault detection during inrush conditions. IEEE Trans Power Deliv 33(6):2965–2977
17. Esponda H, Vázquez E, Andrade MA, Johnson BK (2019) A setting-free differential protection for power transformers based on second central moment. IEEE Trans Power Deliv 34(2):750–759
18. Jiao Z, Li Z (2018) Novel magnetization hysteresis-based power-transformer protection algorithm. IEEE Trans Power Deliv 33(5):2562–2570
19. Farzin N, Vakilian M, Hajipour E (2019) Transformer turn-to-turn fault protection based on fault-related incremental currents. IEEE Trans Power Deliv 34(2):700–709
20. Cai Y, Liu X, Zhang Y, Cai Z (2018) Hierarchical ensemble of extreme learning machine. Pattern Recog Lett 116:101–106
21. PSCAD Research Center (2005) EMTDC-transient analysis for PSCAD power system simulation. Winnipeg, MB, Canada
22. Tripathy M, Maheshwari RP, Verma HK (2007) Probabilistic neural-network-based protection of power transformer. IET Electr Power Appl 1(5):793–798
23. Yang X, Chen W, Li A, Yang C, Xie Z, Dong H (2019) BA-PNN-based methods for power transformer fault diagnosis. Adv Eng Inf 39:178–185
24. Tripathy M, Maheshwari RP, Verma HK (2010) Power transformer differential protection based on optimal probabilistic neural network. IEEE Trans Power Deliv 25(1):102–112
25. Behera S, Tripathy M, Satapathy JK (2016) A novel approach for voltage secure operation using probabilistic neural network in transmission network. J Electr. Syst Inf Technol 3(1):141–150

26. Shah A, Bhalja B (2011) Application of support vector machine for digital protection of power transformer. In: Proceedings of 2011 annual IEEE India conference on engineering sustainable solutions. INDICON-2011, 2011
27. Chothani NG, Bhalja BR, Parikh UB (2012) Development of a new bus zone identification algorithm using support vector machine. IET Gener Transm Distrib 6(7):710–718
28. Chothani N, Bhalja B, Parikh U (2011) New fault zone identification scheme for busbar using support vector machine. Gener Transm Distrib IET 5:1073–1079
29. Li P, Hastie TJ, Church KW (2006) Very sparse random projections. In: Proceedings of the 12th ACM SIGKDD international conference on knowledge discovery and data mining, pp 287–296
30. Ozgonenel O, Karagol S (2014) Transformer differential protection using wavelet transform. Electr Power Syst Res 114:60–67
31. Osman AH, Malik OP (2004) Protection of parallel transmission lines using wavelet transform. IEEE Trans Power Deliv 19(1):49–55
32. Mohanty SR, Pradhan AK, Routray A (2008) A cumulative sum-based fault detector for power system relaying application. IEEE Trans Power Deliv 23(1):79–86
33. Shiddieqy HA, Hariadi FI, Adiono T (2018) Effect of sampling variation in accuracy for fault transmission line classification application based on convolutional neural network. In 2018 International symposium on electronics and smart devices (ISESD), pp 1–3
34. Izykowski J (2011) Power system faults, renewable energy systems. Wrocław University of Technology, Wroclaw

Chapter 7
Real-Time Monitoring and Adaptive Protection of Power Transformer

The power transformer is one of the most important equipment in the grid to reliably and efficiently transmit power to the consumers. Asset management and protection are the best concepts for enlargement of transformer lifespan as well as to increase the grid reliability. This article presents the electrical and non-electrical parameter based power transformer monitoring and protection. Various data such as core flux, age of the asset, heat generation, current harmonics, and temperature are monitored in real-time and process it accordingly to enhance the working capability of the transformer. The proposed scheme is successfully tested on a 15 kVA laboratory transformer using the Arm CORTEX-M4 processor. A Fitness Function (F_f) is estimated from the collected data to examine the working condition of the transformer. Moreover, voltage, current, and power-based inrush detection, as well as Adaptive Power Differential Protection (APDP), are applied to protect the transformer against fault. The hardware implementation and result validation prove the effectiveness of the proposed scheme to enhance the reliability of the distribution grid.

7.1 Literature Reviewed

Trending development in the power system due to several benefits of smart grid and technology nowadays, it is required to change the criteria of protective schemes with self-healing feature. For improvement in overall monitoring and protection of the transformer, it is necessary to analyze all the parameters. Having its self-importance and complexity due to nonlinear magnetizing core characteristics with different voltage levels, transformer protection proves its significance [1]. PLC based transformer cooling control system is applied by intelligent means [2]. Online condition monitoring for distribution transformer is elaborated and discussed in [3]; however, many schemes are lacking the protection criteria in a combination of conditioning monitoring. IEEE has guided the assessment and reconditioning of oil-immersed

D. Patel and N. Chothani, *Digital Protective Schemes for Power Transformer*, Power Systems, https://doi.org/10.1007/978-981-15-6763-6_7

transformers [4, 5]. Time-based maintenance is nowadays replaced by condition-based maintenance as a part of a smart transformer to improve life and reliability with the help of leakage current and partial discharge sensors [6].

Transformer asset management is popularly known as conditioning monitoring and controlling. Dissolved Gas Analyzer (DGA) facilitates to identify transformer conditions. Even, a dissolved gas sensor with the capabilities of multiple gas measuring techniques gives a better prediction for failure possibilities [7]. Fuzzy Logic (FL) based health index is calculated for oil-immersed transformer with real field data [8]. Same, FL based transformer asset management with considering DGA, temperature, etc. is elaborated in [9]. Based on the oil insulation test and FL model decision, a prediction is carried out for the remaining operational life of the transformer [10]. Based on various uncertainty and conflict information, FL is used to evaluate transformer health conditions [11]. Conditional monitoring based transformer asset management greatly increases diagnostic accuracy [12]. The transformer risk index is judged through the asset management plan [13] with optimal physical asset management. To reduce maintenance charges, maintenance strategy is planned based on the evaluation of the life cycle, equipment cost, overhauling time, and repairing cost [14]. Statistical calculations based gradient vector angle of the differential current [15] involved as protective schemes of a transformer. Voltage and the current ratio [16] based scheme is successfully integrated for inrush and fault discrimination. Power Differential Protection (PDP) [17] is a new era of protection scheme which successfully implemented wide area protection with large contingencies. Even PDP based transmission line [18] protection is also well-known among the researchers. On the other hand, the Current Differential Protection (CDP) scheme required phase compensation [19]. Moreover, in the CDP scheme, fundamental components of the current should be extracted to measure the magnitude of the current, and phase angle should be extracted to measure phase difference separately [20]. So, computational complexity will be an increase in the CDP scheme compared to the PDP scheme. One more tragedy can be counted as only current acts as a dominant quantity in the CDP scheme while no other quantities are involved in the CDP scheme [21].

This article describes real-time monitoring of the smart grid transformer by assessing the Fitness Function (F_f). Continuous monitoring of the transformer is achieved by estimating F_f from various parameters like current, voltage, power winding temperature, harmonics, frequency, etc. Breaching the limit of F_f will lead to notify the person at work and will decide isolation of the asset based on its severity of breaching the limit. Moreover, an adaptive power differential algorithm is proposed to protect the transformer against hazardous fault events. Hardware implementation of conditioning monitoring and protection scheme presented here proves the efficacy to improve the performance of the grid-tied transformer.

7.2 Proposed Technique

Figure 7.1 presents a comprehensive diagram for real-time condition monitoring and protection of power transformer. Various data are collected through data loggers to display the collected data and record it for future analysis. The suggested scheme is tested on hardware using the CORTEX M4 processor available in the laboratory. If violation of any considered parameter occurs concerning the predefined limit then estimated F_f provides information to monitoring and protective scheme. With the help of such information, preventive action is carried out to retain the transformer in service for a longer time. Along with monitoring, an Adaptive Power Differential Protection (APDP) scheme is also proposed which will take care of the transformer against internal faults. Both, online monitoring and adaptive power differential protection techniques are executed simultaneously in the hardware processor to enhance the reliability of the transformer in the smart grid.

7.2.1 *Condition Monitoring of Transformer*

Condition monitoring of the transformer is performed by considering certain parameters [3]. We have considered selected parameters like Magnetic Unbalance (MUB), Winding Temperature (WT), Aging Factor, Insulation degradation [22], current

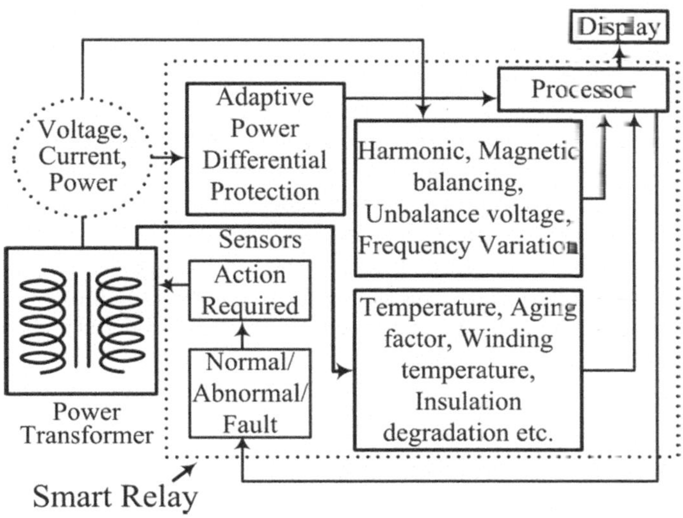

Fig. 7.1 Generalized schematic diagram for transformer monitoring and protection

Table 7.1 Parameters and respective weight factors for the defined fitness function

S. No.	Parameters considered	Score (Si)				Wfi
		4	3	2	1	
1	MUB	0–0.2	0.21–0.4	0.41–0.6	>0.6	3
2	WT	65–70	71–80	81–90	>90	1
3	Aging factor	0.1–0.2	0.21–0.3	0.31–0.4	>0.4	4
4	Insulation degradation	0–0.2	0.21–0.4	0.41–0.6	>0.6	2
5	Current harmonics (%THD)	0–5	6–20	21–40	>40	2
6	Winding deformation	0.0005–0.005	0.0051–0.05	0.051–0.1	>0.1	3
7	Total heat generation	65–70	71–80	81–90	>90	1

harmonics, Winding deformation, Heat generation (total). After acquiring the real-time data, an actual value of said parameters is estimated in the CORTEX M4 processor using a set of equations [3].

Later, a fitness function (F_f) is defined duly considering all effect, as

$$F_f = \frac{1}{S_{\max}} * \frac{\sum_{i=1}^{j} S_i * Wf_i}{\sum_{i=1}^{j} Wf_i} \tag{7.1}$$

where j is the number of parameters, S_i is the score of parameters, $S_{\max}$ is the maximum score of the parameter, and Wf_i is the Weight factor of each parameter. Here, different parameters have different values and units, hence score is defined here as an index which is assigned for a particular range of values, e.g. oil level is in terms of percentage and the winding temperature is in terms of Celsius. Also, the weight factor is assigned based on the dominancy of the parameter i.e. the change in the parameter that highly affects the transformer is assigned '1' weight factor and consecutively in decreasing order up to '4'. One can change the assigned weight factor (Wf_i) and score (S_i) based on their requirements for monitoring of the transformer. Table 7.1 shows the score and weight factors for considered parameters that are acquired for real-time test setup.

7.3 Transformer Protection Approach

The protection of the transformer is significantly important for the reliability of the supply and healthy operation of the power system. Here, real-time monitoring in conjunction with the protection of the transformer is proposed to enhance continuity

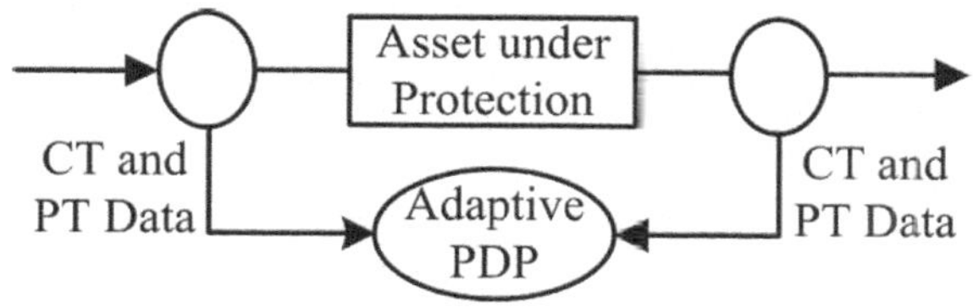

Fig. 7.2 Proposed Adaptive Power Differential Protection (APDP) scheme

of supply in the grid. The acquired parameters used for monitoring are considered for the protection and hence it gives economic operation by eliminating the cost of extra peripheral devices.

In this article, an Adaptive Power Differential Protection (APDP) for the transformer considering CT saturation is proposed as shown in Fig. 7.2. The scheme presented here is more or less similar to that of Current Differential Protection (CDP). However, this scheme offer advantages of adaptive characteristic and reliable operation compare to CDP. The proposed APDP scheme is based on a calculation of the average power of all three phases on both sides of a transformer. This does not require estimation of the fundamental current/voltage magnitude and angles as required in the CDP scheme, phase-wise. Moreover, the computational steps and summation logics are also reduced in the APDP scheme. Here, the input power should be the summation of the output power and losses that occurred in the transformer. If the difference between the powers measured from both the end exceeds the settled threshold then the interpretation is made as an internal fault in the asset and the relay will issue a trip signal.

Also, a novel logic of "Voltage Equality" is proposed which can effectively identify the existence of a fault at the time of inrush condition (transformer energization). The Power Transformer (PT) ratio of both sides of the transformer is taken such that it gives the same output voltage despite any transformation ratio. It is observed that at the time of inrush condition, voltages acquired from both sides of PTs are the same i.e. $V_p = V_s$. This is because of an instant when the primary is energized (under healthy condition) at the same time the secondary of the transformer reflects the desired voltage.

On the other hand, if the case is of faulting the transformer energization, the output of both side PTs will not be equal i.e. $V_p \neq V_s$ and hence there may be the persistence of fault during inrush condition. In the processor, a comparator (logic) is used to compare primary and secondary side PTs voltages for identification of inrush or fault during inrush.

Further discrimination of internal and external fault is carried out by percentage biased adaptive power differential characteristics. As shown in Fig. 7.3, the characteristics are drawn between differential power (P_d) and restraining power (P_r) which is the average power ($(P_1 + P_2)/2$) where P_1 is the primary side power and P_2 is the secondary side power of the transformer.

Concerning the specifications of the transformer, a 30% slope (K_1) is considered a biased setting for relay operating criteria. Hence, if the differential power exceeds

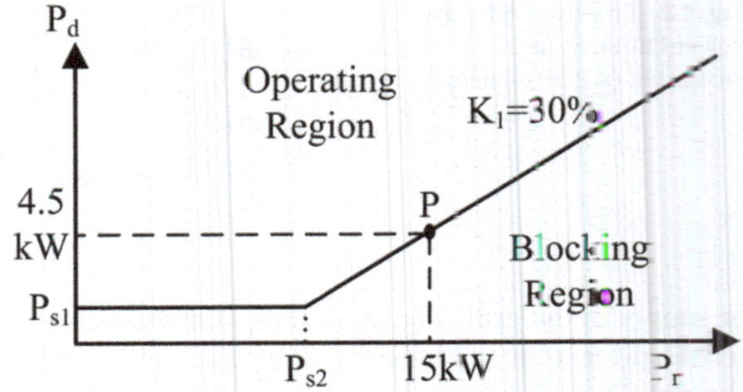

Fig. 7.3 Differential power versus restraining power characteristic

30% of the restraining power (average, in this case), the proposed scheme considers this situation as an internal fault in the transformer.

For the case of an internal fault condition (Fig. 7.3),

(1) When, $P_r < PS_2$ and if, $P_d > PS_1$ then the relay will operate.
(2) When $P_r > PS_2$ and if $P_d > PS_1 + K_1 * P_r$ then the relay will operate.

where K_1 is the initial slope set in the adaptive relay algorithm. PS_1 is the basic differential power setting and PS_2 is the biased power threshold setting.

During a severe fault condition, Current Transformers (CTs) may get saturated [23]. If CTs get saturated during external fault then simple Power Differential Protection (PDP) based scheme may mal-operate. To prevent these types of undesired activities to take place, the PDP scheme is refurbished by adding the adaptive feature. This adaptive characteristic will prevent false tripping during an external fault with CT saturation. The author has also developed an adaptive fault impedance compensation algorithm for the transmission line [24].

Figure 7.4 illustrates the flowchart for the APDP scheme proposed for transformer protection. Initially, currents and voltages of both sides of the transformer are measured with the help of CTs and PTs (same measuring equipment is used for condition monitoring of transformer). As described above, to discriminate against the internal fault and inrush condition at the time of energization, the voltage equality test is performed. If voltages of both the sides of the transformer (V_p & V_s) are not equal then it is considered as the existence of fault and the algorithm further check for the type of fault, conversely, if voltages are equal then the case is of inrush condition.

Further, if the fault condition is detected then the degree of saturation (D_s) of both sides current (I_p & I_s) is measured from Eq. (7.2), which is given below,

$$D_s = 1 - \frac{Saturated\ Current}{Unsaturated\ Current} * 100\% \tag{7.2}$$

where, D_s = degree of saturation.

After that Differential Power (P_d) and Restraining Power (P_r) are estimated using the computational method in the processor. They use digital multiplication of voltage and current samples, acquired from power logger (equipped with A/D converter) [25].

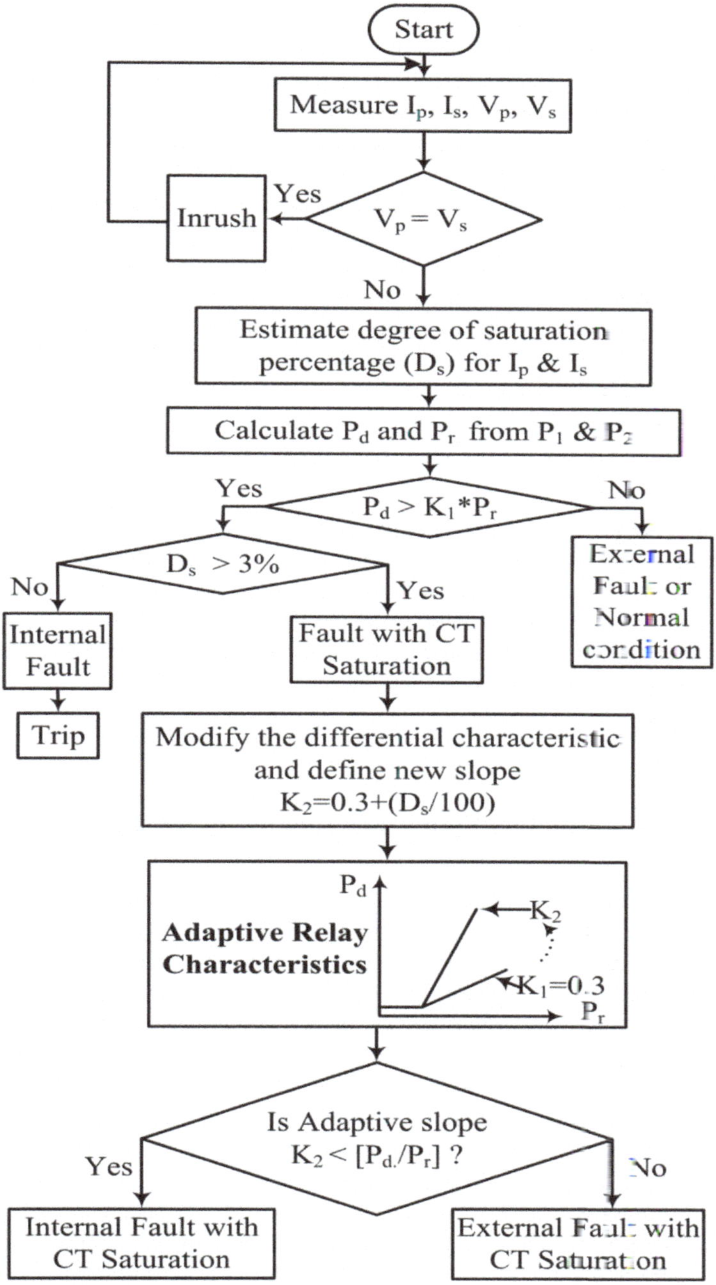

Fig. 7.4 Proposed adaptive PDP based algorithm

If P_d remains less than P_r then the condition is of external fault or normal. But if the P_d exceeds Pr then there may be the presence of internal fault or external fault with CT saturation. Further, if the degree of saturation remains below 3% and already Pd exceeded P_r then it can be understood that the case is of an internal fault condition and the trip signal will be immediately issued to the Circuit Breaker (CB). On the other hand, if the degree of saturation of the currents exceeds 3% then it can be understood as the existence of fault and which should be discriminated properly to prevent false tripping from external fault conditions (Fig. 7.4).

As the saturation level is higher, the necessary action is to be taken to shift the characteristic from the lower slope to higher adaptively, otherwise, the relay will issue a false trip signal. So, the algorithm will now calculate the new slope (K_2) of power differential characteristic with the help of Eq. (7.2), as given in Eq. (7.3)

The new slope is defined as

$$K_2 = 0.3 + \frac{D_s}{100} \tag{7.3}$$

Here, K_2 is the new slope for the APDP scheme.

Analyzing further to identify that the fault is internal or external, the algorithm will again check whether the adaptive slope is less than the ratio of differential power to restraining power. If the new slope is lower than the ratio (P_d/P_r) then it can be concluded that the fault is internal with CT saturation and the trip signal should be generated, otherwise it is decided that external fault with CT saturation occurred (scheme remain inoperative).

7.4 Experimental Test Setup and Result Discussion

The hardware setup is developed in the laboratory to authenticate the proposed real-time monitoring and adaptive power differential protection scheme for the transformer. The snapshot of the developed hardware setup is shown in Fig. 7.5. The transformer considered is a three-phase, 15 kVA, 440/220 V, 50 Hz rated having multiple tapings on both sides. Also, rheostats and inductors are placed before and after the transformer to replicate the effect of the transmission line which is present in the real-time condition. The transformer's primary side is connected with 3-phase, 440-V separate generator, and secondary side is connected to 3-phase 220 V electricity board supply through autotransformer to perfectly create internal fault scenario which takes place in the practical field. CTs are connected on the primary and secondary sides with an appropriate ampere rating. The primary side and secondary side internal faults are generated by connecting 12 A, 18 Ω variable resistors. Additional 250 Ω, a rheostat is inserted in the secondary side of CTs to create a saturation effect during internal as well as an external fault condition. High-resolution DSO and power logger are utilized in hardware to observe and record the current and voltage data during each abnormal condition.

Fig. 7.5 Developed laboratory setup

Various inrush, internal fault and external fault conditions with spectrum analysis and harmonic analysis are carried out on practical aspects. Few selected results are presented here due to space limitations.

7.4.1 *Inrush Condition*

At the time of the transformer energization, the voltage equality test will be carried out as per the algorithm (Fig. 7.4). As shown in Fig. 7.6a when the transformer is switched on from the primary side (closing of CB) at 0.2 s, voltages of the primary and secondary sides of the transformer are equal ($V_p = V_s$) and follow the same waveform pattern as measured from PT secondary. Moreover, calculated RMS values in the processor of these voltages are equal which are shown in Fig. 7.6b. The proposed algorithm recognizes this condition as inrush (healthy energization of the unit) and returns to fetch the next sample.

On the other hand, if the transformer is switched ON in presence of fault within zone then both sides of transformer voltages will be unequal ($V_p \neq V_s$) which can be visible from Fig. 7.6c. The estimated Root Mean Square (RMS) value of these voltages is shown in Fig. 7.6d during faulty transformer energization. If the fault condition identified then the algorithm will further check for the type of fault.

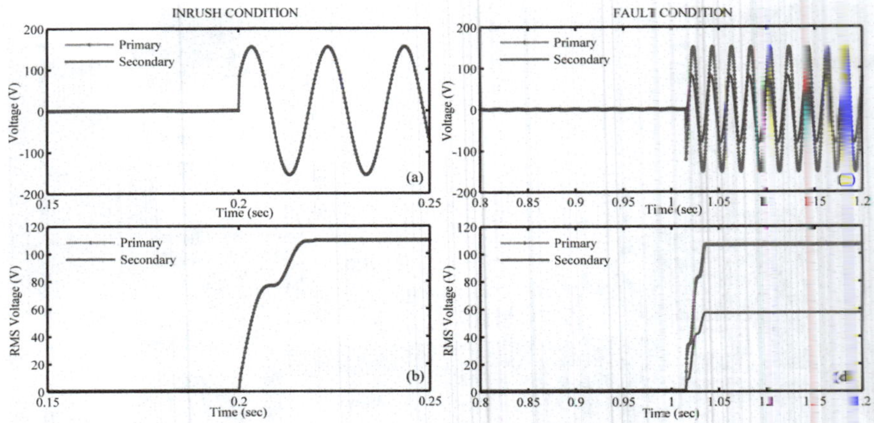

Fig. 7.6 **a** Voltage waveform during inrush. **b** RMS value of voltages during inrush. **c** Voltage waveform during fault. **d** RMS value of voltages during fault

Figure 7.7a shows the recorded waveform of current in DSO at the time of the first energization of the transformer in the laboratory. For cross verification, harmonic analysis and spectrum analysis is carried out as shown in Fig. 7.7b, c, respectively. It is observed that the 2nd harmonic component remains more than 20 percentage compare to fundamental in every phase during inrush condition as shown in Fig. 7.7b [1].

7.4.2 *Internal Fault*

Various single, double, and three-phase faults are carried out on transformer winding through 12 Ω, 18 amps rheostat in a laboratory environment as shown in Fig. 7.8a–c. It is to be noted that for any internal fault whether existing before transformer energization or during operation, the voltage equality test ($V_p = V_s$) must be performed at the very first stage as illustrated in Fig. 7.6c, d. Moreover, during a fault, the relay continuously compares the differential and restraining power concerning set biased slope. As shown in Fig. 7.8d during an internal fault condition, the differential power (P_d) exceeds the restraining power (P_r) times the set slope and consequently, the biased power trajectory falls into the operating zone. Hence, the relay will issue a trip command to the circuit breaker.

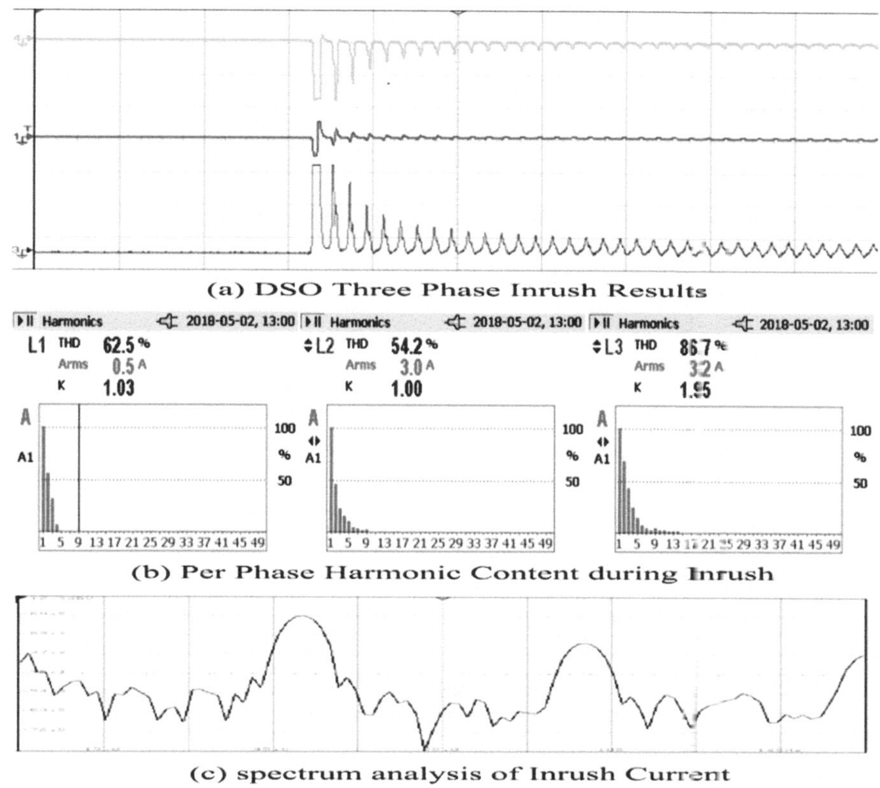

Fig. 7.7 Inrush condition. **a** Three phase inrush currents waveform. **b** Per phase harmonic during inrush. **c** Spectrum analysis during inrush

7.4.3 External Fault or Normal Condition

If we consider the worst case of one CT saturation during an external fault condition, we will get distorted waveforms from CT. This distorted waveform will misguide power differential relay logic to issue trip signal. An adaptive feature has been added in the PDP scheme to tackle the CT saturation during an external fault (Fig. 7.4). An external fault with CT saturation is created on the secondary side of the transformer to check the practicability of the algorithm. A deliberate resistance is inserted in the secondary of one of the CTs to put it into a saturation state. Figure 7.9a–c shows the waveform of CT secondary currents and a comparison of voltages from both sides of the transformer to be protected.

In the case of an external fault condition, the relay should identify the fault outside the protective zone and remains inoperative. At the instant when fault applied in the system, the differential power (P_d) will not exceed the percentage of restraining

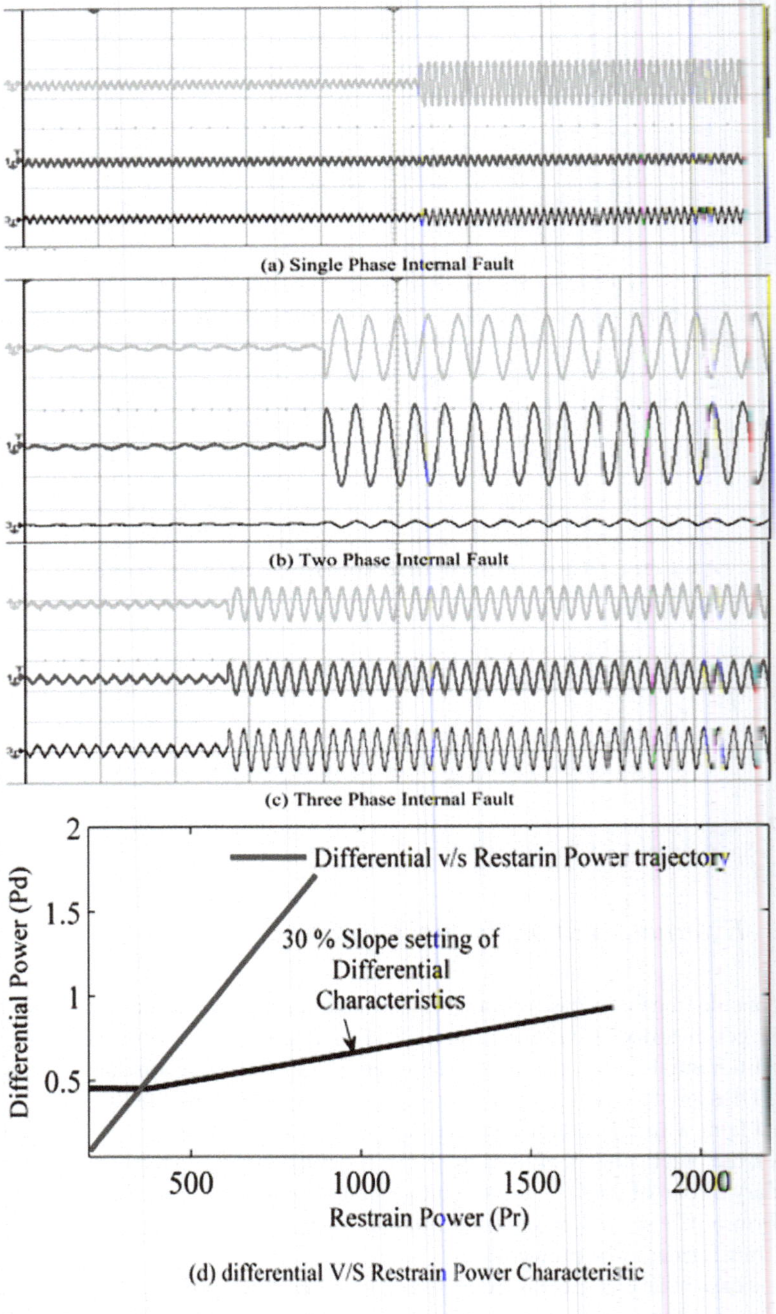

Fig. 7.8 Internal fault conditions

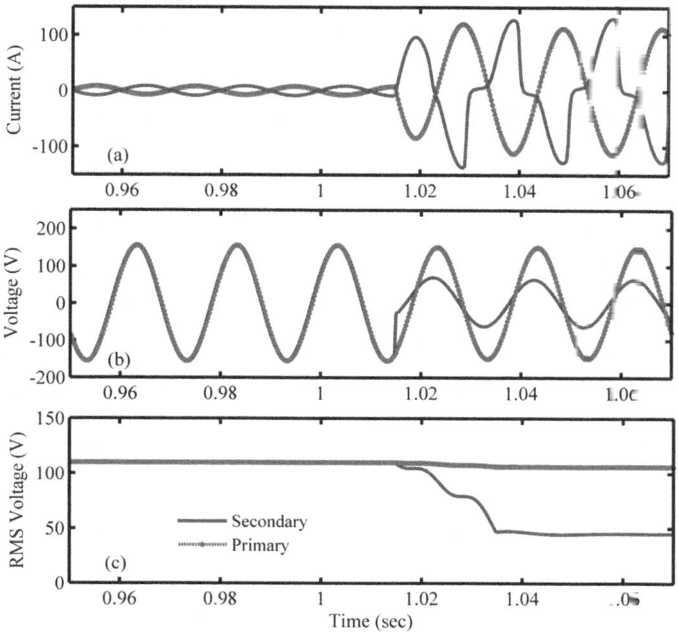

Fig. 7.9 External condition. **a** Current waveform. **b** Voltage waveform. **c** RMS value of voltages

power (P_r) and hence the differential characteristic will remain sufficient below the operating region as shown in Fig. 7.10a. Thus, the proposed scheme will not issue the trip signal.

As per Eq. 7.2, the level of CT saturation is estimated at 15%. The slope of biased characteristic will shift up by 15% as per the Eq. (7.3), hence new slope will be 45% (Fig. 7.10b), which is calculated in the Arithmetic and Logical Unit (ALU) unit of the dedicated processor. The calculated value of P_d remains well below the 45% of P_r, thus adaptive slope prevents false operation of the relay and makes the system more reliable.

7.5 Monitoring of Other Transformer Conditions

The Fitness function (F_f) as described in Eq. (7.1) is estimated for the degradation of various parameter combinations. Figure 7.11a shows the plot of a combination of various parameter variations v/s calculated fitness function. The change in temperature, efficiency, and losses as a function of load variation are also estimated and shown in Fig. 7.11b–d respectively. Table 7.2 and Fig. 7.11a shows the change in F_f for alteration in magnetic unbalance, winding temperature, harmonic content in current and overloading conditions.

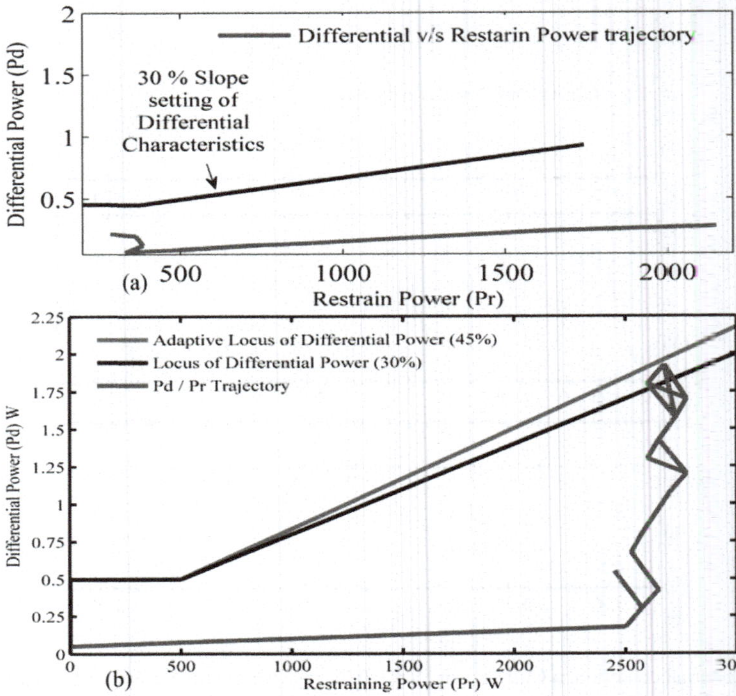

Fig. 7.10 Differential versus restraining power characteristic during external fault condition, **a** without CT saturation, **b** with CT saturation

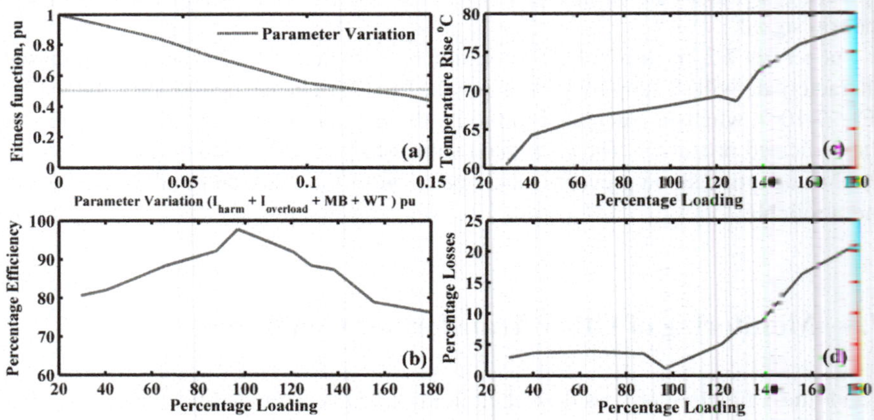

Fig. 7.11 **a** Parameter variation versus fitness function, **b** loading versus efficiency, **c** loading versus temperature and **d** loading versus losses

Table 7.2 Fitness function (F_f) for change in transformer parameter

Parameters variation in pu ($\text{MUB} + \text{WT} + \text{I}_{\text{harm}} + \text{I}_{\text{overload}}$)	F_f (pu)	Grade
0	1	Healthy
0.02	0.92	Healthy
0.04	0.84	Moderate
0.06	0.73	Moderate
0.08	0.64	Poor
0.1	0.55	Poor
0.14	0.47	Worst
0.16	0.4	Worst
0.17	0.37	Worst

The health of the transformer is defined here as a grade, based upon the value of fitness function (F_f). Here we have taken ranges of F_f to reflect its health as, $1 \geq F_f > 0.9$ = Healthy, $0.9 \geq F_f > 0.7$ = Moderate, $0.7 \geq F_f > 0.5$ = Poor and for $0.5 \geq F_f$ = Worst. When calculated F_f enters into the range of poor condition, the proposed scheme provides alarm and in the worst case, it will trip the circuit breakers of the transformer.

An example is elaborated here which adds more light to estimate Fitness function (F_f),

S. No.	Parameters considered	Considered value (pu)	Score (S_i)	*Wfi*
1	MUB	0.2	4	3
2	WT	71	3	1
3	Aging factor	0.3	3	4
4	Insulation degradation	0.4	3	2
5	Current harmonics (%THD)	5	4	2
6	Winding deformation	0.0005	4	3
7	Total heat generation	75	3	1

Here, as per the Eq. (7.1),

$$F_f = \frac{1}{S_{\max}} \cdot \frac{\sum\limits_{i=1}^{j} S_i \cdot Wf_i}{\sum_{i=1}^{j} Wf_i} = \frac{1}{4} \cdot \frac{\sum\limits_{i=1}^{7} S_i \cdot Wf_i}{\sum_{i=1}^{7} W_f}$$

$$= \frac{1}{4} \cdot \frac{(12+3+12+6+8+12+3)}{(3+1+4+2+2+3+1)} = \frac{1}{4} \cdot \frac{(56)}{(16)} = 0.87$$

Hence, the estimated F_f value of the transformer is 0.5. Based on this value, the algorithm will decide on the monitoring of the transformer.

7.6 Summary

This article presents real-time monitoring and protection of power transformer connected in a smart grid. Fitness function (F_f) is derived from the various parameters of the selected transformer. Based on estimated Ff from the collected real-time data for different operating conditions of the transformer, the algorithm provides alarm or trip command. Also, real-time monitoring can display the condition of the transformer from healthy to a worst-case as shown in Table 7.2. Thus, the proposed online condition monitoring will eliminate unnecessary maintenance required for the transformer as scheduled maintenance can be replaced by necessary maintenance. The scheme developed here is generalized; one can modify the parameter based on the requirement. Moreover, an adaptive power differential protection (APDP) scheme is also presented in combination with condition monitoring. The proposed APDP scheme successfully identifies inrush conditions, internal fault, and external fault for the transformer to be protected. The developed approach adaptively modifies power differential relay characteristics during the saturation period of CTs. It is observed that the suggested scheme operates only during internal faults, and remains stable during all external faults, normal load, and inrush condition. The proposed combined real-time monitoring and APDP based protective scheme are successfully implemented on a prototype in a laboratory environment on 15kVA transformer. The results discussed here to justify the combination of the condition monitoring and APDP scheme. Thus, the suggested scheme can be efficiently employed as complete protection of any transformer at different voltage levels.

7.7 Published Article Based on This Work

M. B. Raichura, N. G. Chothani, D. D. Patel, "Real-Time Monitoring Protection of Power Transformer to Enhance Smart Grid Reliability," Electr. Control Commun. Eng., vol. 15, no. 2, pp. 104–112, 2019.

References

1. Bhalja B, Maheshwari RP, Chothani NG (2017) Protection and switchgear, 2nd edn. Oxford University Press, New Delhi, India
2. Pai S, Bansal N, Desai K, Doshi A, Moharkar D, Pathare M (2017) Intelligent PLC based transformer cooling control system. In: 2017 international conference on nascent technologies in engineering (ICNTE), pp 1–6
3. Ballal MS, Jaiswal GC, Tutkane DR, Venikar PA, Mishra MK, Suryawanshi HM (2017) Online condition monitoring system for substation and service transformers. IET Electr Power Appl 11(7):1187–1195

4. Cheng Y, Hu T, Chang W, Bi J (2016) Experiments on the multi-functional current sensor for condition detection of transformer bushings. In: 2016 IEEE electrical insulation conference (EIC), pp 17–20
5. IEEE guide for evaluation and reconditioning of liquid immersed power transformers (2007). IEEE Std C57.140-2006, pp 1–73
6. Najdenkoski K, Rafajlovski G, Dimcev V (2007) Thermal aging of distribution transformers according to IEEE and IEC standards. In: 2007 IEEE power engineering society general meeting, pp 1–5
7. Ceferin S, Janc G, Toro Z, Kastelic T, Pranikar B (2017) Power transformer monitoring systems for better asset management. CIRED Open Access Proc J 2017(1):395–399
8. Abu-Elanien AEB, Salama MMA, Ibrahim M (2012) Calculation of a health index for oil-immersed transformers rated under 69 kV using fuzzy logic. IEEE Trans Power Deliv 27(4):2029–2036
9. Hossam-Eldin AA, Refaey M, Ramadan H (2017) New approach to power transformer asset management and life assessment using fuzzy logic techniques. In: 2017 nineteenth international middle east power systems conference (MEPCON), pp 901–908
10. Bakar NA, Abu-Siada A (2016) Fuzzy logic approach for transformer remnant life prediction and asset management decision. IEEE Trans Dielectr Electr Insul 23(5) 3199–3208
11. Sun L, Ma Z, Shang Y, Liu Y, Yuan H, Wu G (2016) Research on multi-attribute decision-making in condition evaluation for power transformer using fuzzy AHP and modified weighted averaging combination. IET Gener Transm Distrib 10(15):3855–3864
12. Zhang X, Gockenbach E (2008) Asset-management of transformers based on condition monitoring and standard diagnosis [feature article]. IEEE Electr Insul Mag 24(4):26–40
13. Medina RD, Morales DX, Toledo MA, Cabrera JB (2017) Power transformer risk index assessment for an asset management plan. In: 2017 CHILEAN conference on electrical, electronics engineering, information and communication technologies (CHILECON), pp 1–7
14. Takahashi T, Okamoto T (2016) Development of asset management support tools for oil-immersed transformer. IEEE Trans Dielectr Electr Insul 23(3):1643–1647
15. De Alencar RJN, Bezerra UH (2013) Power transformer differential protection through gradient of the differential current. J Control Autom Electr Syst 24(1–2):162–173
16. Ali E, Helal A, Desouki H, Shebl K, Abdelkader S, Malik OP (2018) Power transformer differential protection using current and voltage ratios. Electr Power Syst Res 154:140–150
17. Namdari F, Jamali S, Crossley PA (2007) Power differential based wide area protection. Electr Power Syst Res 77(12):1541–1551
18. Taalab AMI, Darwish HA, Ahmed ES (2007) Performance of power differential relay with adaptive setting for line protection. IEEE Trans Power Deliv 22(1):50–58
19. Sevov L, Zhang Z, Voloh I, Cardenas J (2011) Differential protection for power transformers with non-standard phase shifts. In: 2011 64th annual conference for protective relay engineers, pp 301–309
20. Zhang L, Cong W, Xun T, Bai Y (2011) A current differential protection criterion based on amplitude and phase difference of fault current. In: 2011 international conference on advanced power system automation and protection, vol 1, pp 346–350
21. Dambhare S, Soman SA, Chandorkar MC (2009) Adaptive current differential protection schemes for transmission-line protection. IEEE Trans Power Deliv 24(4):1832–1841
22. Lin C, Zhang B, Yuan Y (2010) The aging diagnosis of solid insulation for oil-immersed power transformers and its remaining life prediction. In: 2010 Asia-Pacific power and energy engineering conference, pp 1–3
23. Chothani NG, Bhalja BR (2014) New Algorithm for current transformer saturation detection and compensation based on derivatives of secondary currents and Newton's backward difference formulae. IET Gener Transm Distrib 8(5):841–850

24. Patel UJ, Chothani NG, Bhatt PJ (2018) Adaptive quadrilateral distance relaying scheme for fault impedance compensation. Electr Control Commun Eng 14(1):58–70
25. Ramos H, Pereira J, Postolache O, Girao P (2002) Minimizing errors due to non-simultaneous sampling of voltage and current in digital power measurement systems. In: Proceedings of the 12th IMEKO TC4 international symposium electrical measurements and instrumentation, 25–27 Sept 2002, pp. 307–310 (Part–1)

Conclusion

Achieved deeper understanding with the practical analysis is a valuable way for measurement and verification of research work. This book mainly presents algorithms for discrimination between internal fault and abnormal conditions in a power transformer. Abnormal conditions like inrush, over-excitation conditions & overloading conditions. Survey of the various methodology and concepts of transformer protection is carried out with proper relevant background, the actual requirement of a field, past events, and current scenarios with consideration of future requirements based on many research articles published in the last 30 years.

One fraction of the book presents a new algorithm for the detection and compensation of CT saturation conditions in the power system. The algorithm is based on a saturation detection index which is obtained using five-point Newton's backward difference formulas. The proposed algorithm is also validated using various CT saturation cases generated in the laboratory environment. Also, based on the comparative evaluation, the performance of the proposed scheme is found to be superior compare to the existing schemes.

Another one description presents a new scheme for the transformer protection based on an average angle of 2nd order derivative of differential current for inrush detection and further discrimination of fault is carried out based on percentage biased differential combined with phase angle comparison between primary and secondary current. The algorithm is developed using the MFCDFT filter to estimate the magnitude and phase angle of current signals. Moreover, the algorithm is authenticated on hardware setup developed in a laboratory environment. One of the advantages of this scheme is a minimum statistical computation.

Further, one part of this book presents adaptive protection of distribution transformer based on the percentage biased differentials principle including a saturation detection method. The algorithm based on CT saturation evaluation and differential principle successfully discriminates against internal fault and external fault. The algorithm is designed using FCDFT and third-order derivative of CT secondary current (saturation detection). The developed approach adaptively modifies differential relay characteristics during the saturation period of CTs.

D. Patel and N. Chothani, *Digital Protective Schemes for Power Transformer*, Power Systems, https://doi.org/10.1007/978-981-15-6763-6

Further two parts of this book present classifier based techniques to avoid maloperation in relaying schemes. So, power system protective schemes must be equipped with appropriate means to correctly identify various abnormal conditions. A Support Vector Machine (SVM) based algorithm is proposed to distinguish various operating conditions. However, a shortfall of SVM/PNN is overcome by RVM and HE-ELM techniques. An RVM based classifier scheme is proposed to discriminate against internal fault, external fault, and other abnormal conditions in a power transformer. The proposed RVM based classifier scheme is compared with existing SVM and PNN based classifier method with observing higher efficiency and require less time to classify faults and inrush current in transformer protection. To check the feasibility of the proposed scheme, hardware-based fault data are generated in the laboratory. Also, HE-ELM based new protective scheme for the transformer in-zone and out of zone fault classification. It turns out from result analysis that HE-ELM outperforms than other classifier techniques like SVM, PNN and ELM.

The last editorial presents real-time monitoring and protection of power transformer connected in a smart grid based on estimated F_f from the collected real-time data for different operating conditions of the transformer, the algorithm provides alarm or trip command. Moreover, an adaptive power differential protection (APDP) scheme is also presented in combination with condition monitoring. The proposed APDP scheme successfully identifies inrush conditions, internal fault, and external fault for the transformer to be protected. The developed approach adaptively modifies power differential relay characteristics during the saturation period of CTs. The proposed combined real-time monitoring and APDP based protective scheme are successfully implemented on a prototype in a laboratory environment.

The result analysis carried out in this work proves the efficiency of the proposed schemes concerning other magnitude or pattern recognition based protection schemes in terms of the accuracy, computational simplicity, and resistance to external disturbances and also generalized for all system parameters.

Future Scope

The author has concluded the techniques based on FFT, DFT, MFCDFT, and HE-ELMRVM/SVM techniques, the classifier techniques like SVM/RVM/HE-ELM are most efficient techniques with having some constraint and limitation of online training and testing issues with grid. A feasibility test is carried out on laboratory prototype work but in the real field, a requirement of training data under various test conditions generates obstacles for classifier techniques also. Most of the cases are incorporated in this research are based on PSCAD simulation and also on hardware.

For the implementation of the algorithm in real field DSP or CORTEX M4 and for capturing current and voltage, current and voltage sensor cards are proposed.

However, some major points out for future scopes are as under

- Day by day so many schemes are developed but the implementation of those schemes online and in the real field itself a major issue with real field data (grid connections) and online updating relaying schemes.
- Implement proposed schemes in a real field (it means in electricity board) which should be cost-effective as well as easy to implement with less maintenance required. Some of the protection schemes though most effective but are complicated and hence required a skilled person at the field to handle it, rather than that protection scheme should be simple which results in easy understanding and implementation in the existing system.
- Along with the protection of the transformer, a monitoring scheme should be developed which can take care end to end operation of the transformer in any situation.
- Moreover, one can develop peripheral communication of the transformer protection scheme (relay) as per the IEC 61850 to communicate with another relay available in a substation.
- The testing of the transformer protection algorithm with high speed advanced controller in real life is a challenging task.

Finally, the main motto is to provide a simple and advanced algorithm to improve the transformer protection scheme.